STUDIES IN HISTORY, ECONOMICS AND
PUBLIC LAW

Edited by the
FACULTY OF POLITICAL SCIENCE
OF COLUMBIA UNIVERSITY

NUMBER 364

SCIENCE AND SUPERSTITION IN THE
EIGHTEENTH CENTURY

A STUDY OF THE TREATMENT OF SCIENCE
IN TWO ENCYCLOPEDIAS OF 1725-1750

Chambers' Cyclopedia: London (1728)
Zedler's Universal Lexicon: Leipzig (1732-1750)

BY
PHILIP SHORR

SCIENCE AND SUPERSTITION IN THE EIGHTEENTH CENTURY

A STUDY OF THE TREATMENT OF SCIENCE
IN TWO ENCYCLOPEDIAS OF 1725-1750

Chambers' Cyclopedia: London (1728)
Zedler's Universal Lexicon: Leipzig (1732-1750)

BY

PHILIP SHORR

AMS PRESS, INC.
NEW YORK
1967

Copyright 1932, Columbia University Press
New York

Reprinted 1967
with permission of Columbia University Press

AMS PRESS, INC.
New York, N.Y. 10003

Manufactured in the United States of America

TABLE OF CONTENTS

PAGE

CHAPTER I
Introduction 7-11

CHAPTER II
Chambers' Cyclopedia and the History of Science. . 12-34

CHAPTER III
Zedler's Universal Lexicon and the History of Science 35-73

CHAPTER IV
Conclusion 74-77

Bibliography 78-79

Index 81

CHAPTER I

INTRODUCTION

SHORTLY after the Peace of Nimwegen (1678), Leibnitz addressed two memorials to Louis XIV; one concerning a plan for the invention of a universal language, the other relating to "Precepts for the advancement of the sciences."[1] In this latter memorial Leibnitz expressed the fear that false views in science and the strife of different opinions might bring about the return of the dark ages of ignorance. Therefore he proposed to the Grand Monarch " to extract the quintessence of the best books, to add to them the unwritten observations of the most tried in every profession, and in this manner to build systems of knowledge, based upon experience and demonstration for the further progress of mankind."[2] Realizing the importance of the century in which he lived, he was eager to take stock of all the science that had been amassed. "What century", he asked, "was better fitted [for such a task] than ours, which will be designated in the future as a century of inventions and marvels?"[2a]

The interesting point in this memorial is that in the suggested project can be seen the first proposal in modern times to compile an encyclopedia of science. Encyclopedias of knowledge were not scarce in western Europe. Throughout the middle ages summations of knowledge were published. And in early modern times these medieval efforts were con-

---

[1] Guhrauer, G. E., *G. W. von Leibnitz—eine Biographie* (Breslau, 1842), pp. 336-337.

[2] *Ibid.*, p. 337.

[2a] *Ibid.*, p. 338.

tinued. Yet none of the newer attempts in the seventeenth century paid attention to the new acquisitions in science and mathematics. Thus, Moréri published his *Grand Dictionnaire* in 1674. But its chief interest was geography, history and biography. In 1690 there was published a dictionary of arts and sciences by Furétière; and in 1694 a more extended work appeared from the pen of Thomas Corneille, a member of the French Academy. This was a two-volume work entitled: *Le Dictionnaire des Arts et des Sciences*. But one man could hardly do justice in two volumes to all the sciences. It is not surprising therefore to find it deficient in its treatment of the sciences.[3] In 1697 Bayle published his critical dictionary. But that work too, though epoch-making as a biographical dictionary, and stimulating in its critical and sceptical attitude toward the orthodoxies of the age, did not pay particular attention to scientific men or to the advances made in science.[4] Hence it may be concluded that although a few intellectuals of the seventeenth century like Leibnitz saw the need of an encyclopedia of science, they had by no means succeeded in materializing Leibnitz's dream.

It is difficult to determine to what extent the encyclopedias of the first half of the eighteenth century were an improvement on their seventeenth century predecessors. Suffice it to say, at this point, that in 1721 J. T. Jablonski published an *Allgemeines Lexicon der Wissenschaften u. Künste* in two volumes; in 1728 Ephraim Chambers published the *Cyclopedia of Arts and Sciences* in two volumes; and between 1732 and 1750 there appeared a German encyclopedia of sixty-four volumes entitled *Grosses Vollständiges Universal Lexicon*, the subtitle indicating that arts and sciences were included. One point might be noted. All three include

[3] See for example the brief treatment of chemistry (100 words), botany (60 words) and medicine (1000 words).

[4] Robinson, H., *Bayle the Sceptic* (New York, 1931), p. 141.

science. But the third, though very much like Chambers' in many respects, departs from all previous efforts in encyclopedic compilations in being a cooperative venture rather than the work of one scholar. And yet this was no guarantee, as will be shown later, that science was to be given better treatment than in the shorter one-man compilations.

The question might well be raised here as to the importance of such encyclopedias, as listed above, for the history of culture in general, and the history of science in particular. This writer, for one, contends that for the student of cultural history a survey of an encyclopedia of knowledge of any particular period might prove helpful in evaluating the state of knowledge of the time, for the simple reason that an encyclopedia is, in a sense, the repository of the civilization of its period; provided it is a comprehensive and accurate work, it ought to give a fair view of the intellectual lay of the land. This seems to be particularly true of the two encyclopedias in question. In the first place, except for a number of differences, the two, an English and a German work, are very much alike in their treatment of the items under consideration. If they were merely representative of an individual philosophy or point of view, there would not be that agreement of approach in almost every subject treated. In the second place, both these works served as guides and bases for *L'Encyclopédie*.[5] Furthermore, in England, Chambers' work proved popular enough to call for the printing of five editions between 1728 and 1743, and other editions in later years. As for Zedler's voluminous *Lexicon*, it was only superseded by the more modern work of Ersch and Gruber.[6]

---

[5] Cru, R. L., *Diderot as a Disciple of English Thought* (New York, 1913), pp. 231 *et seq.*, 237 and 282.

[6] Ersch, J. S. und Gruber, J. G., *Allegemeine Encyclopädie der Wissenschaften und Künste* (Leipzig, 1818), preface, p. viii. Its purpose was definitely to fill the place formerly held by Zedler's somewhat outmoded work, or to produce for the German-speaking people an encyclopedia similar to that of Diderot and D'Alembert.

It is with this general aim in view that the writer has attempted an analysis of both Chambers' *Cyclopedia* and Zedler's *Grosses Vollständiges Universal Lexicon*. But also a very particular purpose led him to those eighteenth century works. It is the relation of those compilations to the history of science which is his very special concern. And for the following reason. Historians of very recent years have been very much concerned with the history of science. Much work has been done in the field. And the results of the labors of such men as Duhem, Haskins, Singer and Thorndike have shown that science has a continuity all its own, that even in the middle ages progress in science can be shown; that the revolutionary scientific discoveries of the seventeenth century had their medieval antecedents. Nevertheless, if one " law " of history points to continuity, change cannot be gainsaid. For how, otherwise, are we to explain the element of novelty in our civilization? Despite the truth, therefore, of continuity, modern science does differ radically from its ancient and medieval prototypes. For one thing, it has freed itself almost completely from magic and occult science from which it had gradually developed and with which it was intimately confused and connected for many centuries. For another, it has abandoned medieval empiricism, and with complete singlemindedness, has devoted itself to objective experimentation. General scientific curiosity to discover the nature of things has displaced the medieval curiosity for the marvelous and supernatural, for the discovery of universal panaceas and the like. It is this change in attitude which is of particular interest to the writer. The question to be considered is, when did this change actually take place? Or, when does medieval science end and modern science begin? The usual answer is that the great revolution in the sciences took place in the seventeenth century. It was then that medieval science received its death blow and modern science was ushered in. The historian, however, who still insists on the "law" of continuity is not satisfied with that summary answer. He questions further. To what extent did the revolution of the seventeenth century effect this change? To what extent, he asks, did the discoveries in physics and astronomy lead to the abandonment of medieval notions in those fields? To what extent did the seventeenth century revolution mark the end of occult science, magic and superstition? Or, if it did not, to what extent did modern science, even after the revolution, carry along in its trail pseudo-science and superstition? It is this problem which the writer has set himself, and has thought of finding a partial solution to it, by surveying the treatment of science and its related fields in early modern encyclopedias. His object in the study of the two above mentioned encyclopedias is to discover the extent of change and the extent of continuity of scientific ideas; his object is to evaluate the extent to which the new scientific discoveries were assimilated in the first half of the eighteenth century and the extent to which medieval science was retained. In this task the writer acknowledges his indebtedness to Professor Lynn Thorndike who has shown him the way by his analysis of Diderot's *Encyclopédie*. In short what will be attempted is a similar analysis of Chambers' and Zedler's works from the point of view of the history of science, adding an English and a German to the French example.

CHAPTER II

CHAMBERS' CYCLOPEDIA AND THE HISTORY OF SCIENCE

CHAMBERS' *Cyclopedia,* published in 1728, is largely a systematic compilation of accumulated knowledge. And that is perhaps what a cyclopedia should be. It has no point of view of its own; the lexicographer makes no effort to be any more than a recorder of the accepted ideas and facts of his time. In that respect it differs from *L'Encyclopédie* which was published some two decades later and was at first intended to be little more than a French adaptation of Chambers'. *L'Encyclopédie* is not merely a record of civilization, it is also a critique. Undoubtedly it laid the intellectual foundation for the French Revolution. But in another respect Chambers' *Cyclopedia* may be considered a forerunner of *L'Encyclopédie.* Aside from its attack on Church and State, *L'Encyclopédie* is chiefly concerned with science. "Scientific research is its very essence." [1] In its very title, *Encyclopédie ou dictionnaire raisonné des sciences, des arts et des métiers par la société des gens de lettres,* science comes first. Chambers too is much concerned with science. He is almost willing to compare his age with that of Augustus because of its great progress in experimental science.[2] And his *Cyclopedia* he calls a "*Universal Dictionary of Arts and Sciences.*" To be sure both Diderot and D'Alembert were far greater

---

[1] Thorndike, L., "L'Encyclopédie and the history of science," in *Isis,* VI (1924), p. 363.

[2] In his address to the King Chambers says: "The time is now at hand when we are no longer to envy Rome her Augustus and Augustan Age; but Rome in her turn shall envy ours. . . . Numerous presages give us room to expect that . . . what Greece was under Alexander, and Rome under Augustus Caesar, Britain shall be under George and Caroline."

men than Chambers. Diderot had made his contribution in the field of medicine and biology, and D'Alembert in mathematical science and celestial mechanics.[3] But if Chambers was not an original thinker, he showed a great interest in science in attempting single-handed so tremendous a task, and completing it more or less competently, and also later in collaborating with John Martyns in an abridged translation of the philosophical history and memoirs of the Royal Academy of Science at Paris.[4]

It might be said at the very outset that Chambers [5] treats many subjects as they would be treated in an encyclopedic dictionary, rather than an encyclopedia. This accounts to a great extent for the lack of a point of view. It is very apparent especially in his treatment of the subject of history.[6] Here is merely set down a short definition and classification of history into ancient and modern, sacred and profane, universal and particular. Whatever philosophy of history may have been current at the time is not revealed in his account. Chambers' own philosophy can only be inferred from a brief statement in his address to the King that " there is a time for every nation to arrive at its height, and the uppermost place on the terrestrial ball is held successively by several states." But that is of no importance at all in his treatment of the sciences, except that there is a tendency to describe in great

[3] Thorndike, L., *op. cit.*, p. 365.

[4] *Philosophical History and Memoirs of the Royal Academy of Science at Paris*, trans. by John Martens and E. Chambers, 5 vols.

[5] A brief note on Chambers in *Dictionary of National Biography* tells us that he translated from the French of Jean Debreuil the *Practice of Perspective* of which a fourth edition appeared in 1765; that he collaborated with the botanist John Martyns in an abridged translation of the *Philosophical History and Memoirs of the Royal Academy of Science at Paris (1745)*; and that he was an avowed free thinker. See also Cru, R. L., *Diderot and English Thought*, p. 231. According to Larousse in art. "*Encyclopédie*," Chambers was buried in Westminister.

[6] See article, "History."

detail modern mechanical inventions, an example which was to be followed by *L'Encyclopédie*. As for conceiving history as an exact study or science, that is nowhere apparent. In general it may be said that his historical accounts of the sciences are never very complete and always contain gaps.

To begin with his treatment of astronomy, his historical account is rather brief. Passing over very hurriedly the contributions of the Greeks, he makes the dubious statement that Pythagoras " was the first among the Europeans who taught that the earth and planets turn around the sun which stands immovable in the center; that the diurnal motion of the sun and fixed stars was not real but apparent, arising from the earth's motion around its own axis ". It is usually held that Pythagoras considered the earth the center of the universe, allowing the sun, moon and planets to build circular paths around it. It was Aristarchus who set forth the heliocentric theory; but Chambers considers it sufficient to mention his name only. Of Ptolemy Chambers merely remarks that he was " the great Alexandrian astronomer ". He does make note of an interesting fact concerning the Ptolemaic system which seems to indicate that Ptolemy as late as 1728 had not entirely lost his influence. For the " principal assertions of the Ptolemaic system . . . are still adhered to in some Universities where free philosophizing is excluded ".[7]

The history of astronomy during the middle ages is not treated at all. Neither the astronomical and trigonometric tables of Al-Khwarismi,[8] nor the *Sphaera Mundi* of John Sacrobosco find a place in Chambers. The *Sphaera Mundi* was not only a useful book in the thirteenth century but for many centuries thereafter it was the classic textbook in

[7] See article " Ptolemaic System."

[8] Sarton, G., *Introduction to the history of science* (Washington, D. C., 1927), p. 545.

astronomy. Many commentators made note of it, and continued to do so until the end of the seventeenth century.[9] There were as many as fifty-nine editions of it, and a translation in Italian. Surely his name might have been mentioned. Gerard of Cremona, who made accessible many Arabic works in astronomy, meets a similar fate in Chambers' *Cyclopedia*. Another omission in Chambers is one of the most popular works in the middle ages, namely, the *De Scientia Motus Orbis*, an Arabic work of Mashallah (Messahalla) translated into Latin by Gerard of Cremona.[10] Adelard of Bath might have been mentioned. "The work of Adelard of Bath remains comprehensive and fundamental alike with reference to mathematics, astronomy, astrology, and his advocacy of experimental method".[11] But perhaps Chambers ought not to be blamed too severely for neglects of this kind. After all the task he undertook was a colossal job for one man; hence as a popular reference book, whatever history he included was sufficient. Yet when we are compelled to consider this work as a universal dictionary or a cyclopedia of science, it may be criticized for its omissions.

[9] Wolf, R., *Geschichte der Astronomie* (München, 1877), p. 208.
[10] Sarton, G., *op. cit.*, p. 119.
[11] Haskins, C. H., *Studies in the History of Medieval Science* (Cambridge, 1924), p. 119. At the end of the chapter devoted to Adelard of Bath, p. 42, Haskins makes the following significant comment, in a sense explanatory of the above citation: "Adelard occupies a position of peculiar importance in the intellectual history of the Middle Ages. Standing at the point where the traditional knowledge of the Cathedral schools meets the new learning of southern Italy and the Mohammedan east, his attitude was one of personal inquiry and not mere blind receptivity. The first, so far as we know, to assimilate Arabic science in the revival of the twelfth century, to him we owe the introduction of the new Euclid and the new astronomy in the west. Moreover, he was a pioneer in more than the chronological sense. He went out to seek knowledge for himself by travel and exploration, penetrating as far as Sicily and Syria, and, probably, Spain; and he showed a spirit of independent inquiry quite his own."

In the article on astrology we learn that judicial astrology "which pretends to foretell moral events" is superstitious but that natural astrology has sound basis in fact. Judicial astrology, however, had been denounced many times before Chambers. Isidore of Seville [12] in the seventh century, for example, made the distinction between natural and superstitious astrology. Hence there is nothing strikingly sceptical about the position taken. Nor can it be said that this condemnation of judicial astrology is a result of scientific advance. Indeed, the argument advanced against the art, although based on common sense rather than on religious or moral grounds, is still rather old. To cite one passage,

You maintain [says the author quoted by Chambers] that the circumstances of life and death depend on the place and influence of celestial bodies, at the time when the child comes to light; why are we to regard only the stars at his nativity, and not those rather which shone when the foetus was animated in the womb? And why must those others be excluded which presided while the body remained tender and susceptible of the weakest impression during gestation?"

On the other hand, the argument supporting natural astrology seems to indicate that nothing much has been gained from the change from geocentric to heliocentric conceptions. Chambers need only quote Boyle to feel quite safe. Thus Boyle claims that properties of moisture, heat and cold, and the like are wholly dependent on the course, motion and position of the heavenly bodies; that every planet must have its own proper light distinct from any other; and endowed with its (i. e. the planet's) specific powers. The sun not only shines on all planets but raises the motions and properties peculiar to them. Now the light of each planet modifies the reflected rays of the sun which are in return again

---

[12] Hearnshaw and Barker [editors], *Medieval Contributions to Modern Civilisation*, 1921, p. 132 in article "Science" by Singer.

reflected into other parts, particularly the adjacent bodies of the planetary system. The powers of each planet on sublunary things depend upon the angle the planet makes with the sun and upon the distance from the earth.[12a]

As for the moral and physical influence of comets,[13] Chambers must be given credit for abandoning the former, and accepting in a very hesitant manner the latter. As late as 1770 it may be noted that Prof. Charles Gottlieb Semmler, professor of mathematics and physics at Halle, still believed in the moral influence of comets,[14] As for physical influence, Chambers cites Newton[14a] to the effect that comets are absolutely necessary for the conservation of the water and moisture of the planets; from these condensed vapors and exhalations all the moisture, spent in vegetation and putrefaction

[12a] See article "Astrology"; also Boyle, R., *Philosophical Works* (edited by Peter Shaw) (London, 1738), vol. iii, p. 34. The influence of the planets, he says, "appears by undeniable experiments, not only in vegetables but in animals, and that both in acute and chronic distempers; more particularly in lunatic, epileptic, paralytic or lethargic patients." And then on p. 35, "As for the manner wherein the planets transmit their powers, and thereby affect the remoter bodies, 'tis not difficult to apprehend it; for we affirm no virtue or power to flow from the planets, that comes not along with the light as a property thereof."

[13] See article, "Comet".

[14] Wolf, R., *op. cit.*, p. 186.

[14a] See article, "Comet". Also Newton, I., *Principia* (reprinted, Glasgow, 1871), p. 515... " ad conservationem marium & humorum in planetis requiri videntur cometae, ex quorum exhalationibus & vaporibus condensatis quicquid liquoris interram aridam convertitur, continuo suppleri & refici possit . . . Porro suspicor spiritum illum, qui aeris nostri pars minima est sed subtilissima & optima & ad rerum omnium vitam requiritur, ex cometis praecipue venire."

("Comets seem to be necessary for the conservation of seas and fluids in the planets, so that whatever of liquid is converted into dry earth, may be continually supplied and made up from their [i. e. comets] exhalations and condensed vapors . . . I suspect moreover that it is chiefly from the comets that spirit comes which is the smallest but the most subtle and best part of our air, and is essential for the life of all things.")

and turned into dry earth, may be resupplied and recruited. "For this reason," says Chambers, "there may be some ground for the popular belief in the presages of the comets, since the tail of the comet thus intermingled with our atmosphere may produce real changes in the animal and vegetable kingdom ".

On the whole the subject of physics is treated fairly competently and fully, perhaps because so much more progress had been made in that science. A good deal of space is devoted to verbal and pictorial descriptions of mechanical inventions and the like. However, separate phases of physics are not always treated as completely and fully as one might expect in an encyclopedia. The article on light, even for that time, is insufficient. Whereas Chambers in many instances takes the trouble to present views that had been discarded and seems to show a non-partisan attitude, on the matter of light he considers it sufficient to explain in a very detailed manner, and justly so, the Newtonian corpuscular theory,[15] but does not so much as mention Huyghens' undulatory theory. Also in his brief history of the subject there is no reason for omitting Plato's theory of vision when Aristotle's is included. As for the work in the field of light and optics in the middle ages, the cyclopedia is very weak and brief, and even of seventeenth century works only a few names are mentioned. Alhazen and Witelo are mentioned but nothing is said as to their actual contribution. That Alhazen is an important figure in the history of optics cannot be gainsaid. He made an anatomical study of the eye and explained the function of its parts. He believed that vision resulted from emanations from visible objects, and also showed how the eye received cones of rays from each point of the object and not single rays as had been supposed. He also noticed refraction of the atmosphere [15a] and showed how phenomena

---

[15] See article, "Light".   [15a] This was known also to Ptolemy.

of twilight were connected with the height of the atmosphere.[16] Alhazen's work exercised great influence on European thought and was the basis of several commentaries on optics by Witelo (1270) and others.[17] Other important names are mentioned, but no works indicated. Della Porta, for example, is mentioned but nothing is said of his *Magia Naturalis* (1558) in which he described the camera obscura (magic lantern), spectacles, and arrangements of lenses suggestive of the telescope.[18] Leibnitz and his *Principe Générale d'Optique* are mentioned without further comment.[19]

The attitude toward the experimental method is characteristic of the period. We are told that experiments are of the greatest importance in philosophy; that the great advantage of modern over ancient physics is that greater use is made of the experimental method. But Chambers is compelled to note that there were still in his day opponents of experimentation. Experimenters, these opponents claimed, were not true philosophers but laborers. One writer cited believed that these experimenters were only a " tribe of idly curious people whose philosophy consists in making experiments on the gravity of air, the equilibrium of fluids, the loadstone, etc., and yet arrogate to themselves the noble title of philosophers." Another opponent admitted that " philosophy has received very many advantages from makers of experiments ", but complained of their " disingenuity in too often wresting and distorting their experiments and observations to favor some darling theories they had espoused ".

Concerning the treatment of chemistry in the *Cyclopedia*, the question arises to what extent chemistry, by 1728, had been divorced from alchemy and medicine or from the notions

---

[16] Buckley, H., *A Short History of Physics* (London, 1927), p. 62.
[17] Buckley, H., *op. cit.*, p. 62, and Sarton, *op. cit.*, p. 721.
[18] *Ibid.*, p. 62.
[19] See article, " Light ".

held by Paracelsus and other iatrochemists. Robert Boyle had done a good deal to lay the foundations of modern chemistry. In the preliminary discourse to his works, he distinguished very clearly between the medieval study, which concerns itself with medicine and alchemy, and our modern notion of the science.

> I saw that several chymists had by a laudable diligence obtained various productions and hit upon many more phenomena. considerable in their kind, than could well be expected from their narrow principles; but finding the generality of those addicted to Chymistry to have had scarce any view but the preparation of medicine or to improving metals, I was tempted to consider the art not as a physician or an alchymist, but as a philosopher. And with this view I once drew up a scheme for a chymical philosophy, which, I should be glad that any experiments or observations of mine might in any way contribute to complete. . . .
>
> And if men were willing to regard the advancement of philosophy more than their own reputation, it were easy to make them sensible that one of the most considerable services they could do the world is to set themselves diligently to make experiments and collect observations without attempting to establish theories upon them before they have taken notice of all the phenomena that are to be solved.[20]

Now what of such ideas as expressed by Boyle has penetrated into the *Cyclopedia*? In the first place Chambers finds no space to summarize Boyle's ideas on chemistry, but does find room for a short note on his sermons, although that scientist's works were available long before the cyclopedia appeared. Secondly chemistry is still subdivided into " alchymia, metallurgia, chymical pharmacy and chymical philosophy." Alchemy is considered the " sublime part of the art of chymistry," or a higher and more refined kind of

---

[20] Boyle, R., *Philosophical Works* (edited by P. Shaw) 1738, pp. xxvi-xxvii, xviii, vol. i (cited by Meyer, E., *A History of Chemistry*, pp. 103-104).

chemistry employed in the "more mysterious researches of the art". The four elements as taught by Aristotle are accepted. In the article on water as one of these elements, Chambers relies on the authority of Basil Valentine, Paracelsus, Van Helmont, Sendivogius and others who maintained the principle that water is elemental matter. Yet Boyle in his *Skeptical Chymist* (1661) refutes the three men who support Aristotle's notion of the four elements and Paracelsus's theory of three substances comprising all matter.[21] Considering the fact that Boyle is cited on any number of occasions, especially when it is necessary to corroborate some superstitious notion, as in the case of the efficacy of amulets, it is very surprising that Chambers does not take the trouble to set down Boyle's ideas side by side with those of Paracelsus and Aristotle. Somehow one cannot imagine that he did not see or read Boyle's book. But it does seem that Boyle's ideas were perhaps too advanced for his time, for even the French encyclopedists who were imbued far more thoroughly than Chambers with the scientific spirit, do not give a more scientific treatment of the subject; in fact it resembles in every way the treatment found in Chambers.[22]

The article on alchemy [23] retains all the medieval science connected with it, without any change. The great objectives of alchemy are enumerated without a shadow of scepticism. These are (1) the turning of baser metals into gold, to be effected by the philosopher's stone; (2) the finding of the elixir or universal medicine; (3) the discovery of the *alkahest* or universal solvent; (4) and the search for a universal ferment which when applied to any seed will increase its

[21] Kopp, H., *Beiträge zur Geschichte der Chemie* (Brunswick, 1869-75), note p. 166, v. iii.

[22] Thorndike, L., "L'Encyclopédie and the history of science," p. 380.

[23] See article, "Alchymy".

fecundity. That the subject still absorbed the interest of a good many readers of a cyclopedic work is evident from the large amount of space Chambers devotes to alchemy. He is very careful in his historical account of the subject. Then he has comparatively long accounts on the *alkahest* and philosopher's stone. All in all we cannot say that very much scepticism is betrayed in the treatment of alchemy and related topics. And yet the writer wonders whether one is not perhaps too censorious in his criticism of Chambers. Even Boyle was so fond of the idea of the universal solvent, that he would rather have been master of it than of the philosopher's stone.[23a] Moreover, Chambers adds on his own account that a universal solvent is not so absurd as it may seem, for all bodies must have originally arisen from some first matter which was once fluid in form. Thus gold which in its primitive state may have been a heavy fluid, as a result of a strong attraction between its parts afterwards acquired a solid form.

Although botany is still treated as a branch of medicine, it is defined as the science of herbs or that branch of medicine and agriculture which treats of plants whether medicinal or other, their several kinds, forms, virtues and uses. There is a detailed account of the science of botany including all the progress made in the field throughout the seventeenth century.[24] Also in the description of individual plants very few of the old superstitions have survived. In the case of the heliotrope, Chambers says that some have ascribed to it the power of rendering people invisible, like Gyges' ring.[25] It is very doubtful whether Chambers took this view seriously; it would seem rather that he inserted it for the amusement of

[23a] See article, "*Alkahest*". Chambers does not cite any particular work of Boyle in this connection.

[24] See article, "Plants".

[25] See article, "Heliotrope".

his readers. Also in his remarks on the *agnus castus,* he does not omit to tell his readers that it was very famous among the ancients as a specific for the preservation of chastity, and that the Athenian ladies who made profession of chastity lay upon leaves of *agnus castus* during the feast of Ceres. Yet he is quick to add this time that "it is out of present practice." [26] The mistletoe, however, had not yet lost its medieval virtues. It still was most efficacious against epilepsy; it was also prescribed as a cure against apoplexy, lethargies and vertigos. And it was worn about the necks of children to prevent convulsions and the pain of cutting their teeth. Although Peysonnell in 1725 had discovered that coral was a hard skeleton secreted by certain marine animals,[27] Chambers classifies coral as a plant, though he scouts the virtues attributed to it. But as for his error in classification of coral it might be noted that in a translation of Theophrastus's history of stones in 1746, the same mistake is repeated.[28]

In his discussion of precious stones,[29] Chambers has abandoned to a large extent the belief in their supernatural powers. This might be best illustrated in his article on the beryl. He rehearses, to be sure, the wonderful properties attributed to it by the ancients, that it kept people from falling into ambuscades of enemies; that it excited courage in the fearful; and that it cured diseases of the eye and stomach. But then he adds that "it does none of these things now, because people are not simple enough to believe that it has the virtues to do them". And yet this attitude is not always maintained in describing other stones. In the case of the agate he repeats the ancient belief that it was a preservative against the

[26] See article, "*Agnus castus*".
[27] Thorndike, L., *op. cit.*, p. 382.
[28] *Ibid.*, note p. 382.
[29] See articles "adamant", "beryl", "topas", etc.

poisons of vipers, scorpions and spiders, and says not a word to discredit such a notion. Even more glaring an instance is his article on the stone *aetites*, " a tophaceous crustated stone hollow within," as he described it. After enumerating some of its extraordinary magical powers, accredited to it by the ancients, he says, " the use now made of it is to assist women in labor; to which end they fasten it about the knee, it being a tradition that according as it is applied above or below the matrix, it has the faculty of retaining or excluding the child ". But even if the magical virtues of precious stones have been abandoned to some extent, their therapeutic value is still strongly affirmed. In speaking of gems [30] he points out that some people doubt the medicinal virtues of stones. Even so the fragments of stones are still retained in physicians' prescriptions. " When much the greatest part of their traditional qualities are set aside as fabulous, there will still remain some, on as real and well warranted a footing as many of our other medicines ". Next he cites Boyle's treatise on the origin and virtue of gems, summarizing Boyle's thesis that originally gems were in a fluid state, and that many of their general virtues were derived from a mixture with metallic and mineral substances incorporated with them. He then devotes a little more than a column to refuting the arguments of those who deny the medicinal virtue of gems.

During the period 1680-1748, Cartesianism, which meant the supremacy of reason over authority, transformed human thought. Thus Bury in his *Idea of Progress*.[31] The question arises, did that mean the abandonment of old ideas, superstitious and otherwise erroneous, and the adoption of the scientific spirit of inquiry? Or may not this transformation merely have meant the abandonment of some of these

[30] See article, " Gems ".
[31] Bury, J., *The Idea of Progress* (London, 1920), chapter on Fontenelle.

notions, but the retention of a good many of them through a process of rational and pseudo-scientific explanations? After reading many of the articles in the *Cyclopedia,* related to science and allied fields, the writer feels inclined to accept the second view of this great change. Otherwise one would be hard put to it to explain a good many of the ancient beliefs retained in the *Cyclopedia.* In the field of medicine for example, vital, natural and animal functions are given a prominent place;[32] and the four humors,[33] although frowned upon by Chambers, are explained. A very vital problem is in what part of the brain the mind resides, the consensus of opinion being that it is located in that part of the brain where all the nerve fibres end.[34] There is some disagreement, however, as to the exact location of the soul, some believing it to be equally diffused throughout the body, and others saying it influences every part but resides in one part, the sensory. This part Descartes maintained was the pineal gland of the brain, where the nerves terminate.[35] Nor has the notion of "animal spirits" been abandoned.[36] In the brain is found a very subtle and fragrant juice, which is the principal seat of the reasonable soul and is identified with "animal spirits". Whether or not animals have a soul was still a favorite question. The Cartesians denied that animals possessed souls, or had feelings of pleasure or pain.[37]

The article on the brain as well as others would seem to indicate that marvels found an important place in Chambers' *Cyclopedia.* The assertion is made, for instance, that the brain is not absolutely necessary to animal life, and for proof

[32] See article, "Function".
[33] See article, "Humour", also article, "Medicine".
[34] See article, "Sensation".
[35] See article, "Soul", and article, "Brain".
[36] See article "Soul".
[37] See article "Animal".

instances are mentioned where children were born alive without brains. In one instance a child was said to have been delivered at maturity and to have lived for four days not only without a brain but without a head.[38] Other freaks of nature or marvels abound in the *Cyclopedia*. There is, for example, the case of a girl in Germany who for two years vomited toads and lizards; but the most curious of all was an instance where a kitten was bred in the stomach and then vomited up. There are also instances, adds Chambers, of whelps, frogs and other animals that were bred that way.[39] All these freaks of nature were taken from medical authorities. In all fairness to the cyclopedist, it must be added that some of the ancient marvels were rejected. The story of the phoenix [40] that is said to have lived from five to six hundred years is told in all its details but " is discredited by moderns as fabulous ". He also discredits the common tradition that the chameleon [41] lives on air for the reason that experience shows the contrary; after enumerating all the marvels of the chameleon believed in by the ancients, he discredits them as superstitious.

But to return to the treatment of medicine in the *Cyclopedia*. The history of medicine has many lacunae. Although Chambers indicates the importance of the Arabic contributions to medical science, not one of them is mentioned. Nothing is said of the first European medical school at Salerno, the importance of the Montpellier school in the middle ages or the work of Arnald of Villanova or Peter

[38] See article on "Brain".
[39] See article on "Voiding".
[40] See article on "Phoenix". The naturalists according to Chambers hold that the phoenix lives from five to six hundred years. Then it builds itself a funeral pile of wood and aromatic gums, and lights it with the wafting of its wings, thus burying itself. From its ashes arises a worm which in time grows up to be a phoenix.
[41] See article, "Chameleon".

of Abano. Not only are the earlier middle ages neglected but even Paracelsus of the sixteenth century is missing in this historical account. And Paracelsus has been recognized as the most original medical thinker of that century. " While philosophy, alchemy and astronomy were the pillars of his faith, his watchword in practice was experimentation controlled by authoritative literature ".[42] As for Galen, Chambers is somewhat critical. Galen did much harm in the field of medicine because he stressed too much the doctrine of the elements, the cardinal qualities and their degrees, and the four humors, and the " moderns do not pay much attention to such divisions ".[43]

Yet it would be very far from the truth to say that Chambers had discarded Galen and other ancients entirely. A good many past conceptions and theories in medicine are retained. The body, for example, is composed of solids, humors and spirits.[44] Diseases of the fluid are either in the mass of the blood or the spirits; diseases of animal spirits arise either from intermission or retardation of motion. He elaborates in great detail the classification of diseases by Boerhaave,[45] who, according to Allbutt, made no experiments in medicine but " seems to have contented himself with hashing up the partial truths and the entire errors of his time ".[46] Sydenham is another favorite authority of Chambers. But for Sydenham " pathology was summed up in the Hippocratic theory of the concoction of the humors and the subsequent discharge of *materies morbi.*" [47] Chambers cites his

[42] Garrison, F. H., *An Introduction to the History of Medicine* (Philadelphia, 1917), p. 188.
[43] See article, " Galen ".
[44] See article, " Disease ".
[45] See article, " Disease ".
[46] Allbutt, T. C., *British Medical Journal* (London, 1900, ii), p. 1850 (cited by Garrison, *op. cit.*, p. 309.)
[47] Garrison, F. H., *op. cit.*, p. 261.

definition of fever as "a strenuous endeavor of nature to throw off some morbific matter that greatly incomodes the body". Garrison says of him that "he stood off from all medical theorizing and scientific experimentation of his time, disregarded all predecessors save Hippocrates and knew nothing whatever of Vesalius, Harvey, Malpighi or Mayo."[48] Then again the heart and the liver are considered as the seat of "the passions and inclinations".[49] Galen's pulse doctrine, a special pulse for every disease,[50] is still retained despite the work of Sir John Floyer (1649-1734) called the *Physicians' Pulse Watch*.[51] Also phlebotomy is still the favored treatment in both increasing and retarding the circulation of the blood. Thus,

if there be a due strength and elasticity remaining in the solids, phlebotomy will make the remaining blood circulate the faster and become thinner and warmer; but in a plethora resulting from debauch phlebotomy will make the remaining mass circulate slower and become cooler. . . . In the former case a diminution of the resistance in the blood vessels will increase the contractile powers of the vessels and make them beat faster and circulate their contents with greater velocity, but in the latter case a diminution of the quantity of the spirituous blood will lessen the quantity of spirit secreted in the brain . . . and therefore the blood will move slower and become cooler.[52]

In discussing the treatment of specific diseases, Chambers retains a good many absurdities of his own and more remote times. Some of the symptoms of hydrophobia are said to be snarling and barking like a dog; and even the trans-

[48] *Ibid.*, p. 260.
[49] See article, "Chiromancy".
[50] See article, "Pulse".
[51] Garrison, F. H., *op. cit.*, p. 355.
[52] See article, "Phlebotomy".

formation of organic parts of the body to resemble those of a dog. The remedies proposed are identical with those of ancient and medieval writers on the subject. Galen's remedy to feed the patient a spoonful of ashes of either crawfish or river lobsters twice a day for forty days is given as a sure cure in case external precautions had not been taken in time. Others are even more absurd than this, and are only worthy of repetition here to indicate to what extent intelligent men with an apparently scientific outlook can accept the most unreasonable and far-fetched ideas. One is to pluck the feathers from the breach of an old cock, and then apply the cock bare to the bite.[53] If the dog was mad, then the cock will swell and die and the patient will recover. If the cock does not die, then the dog was not mad and there is nothing to worry about. Or roast an onion under the ashes, and apply it to the wound. Another most certain cure is to apply to the bite a hair of the same dog; this has the strange power of attracting poison, thus effecting a cure. Best of all are the data of cures furnished by the Royal Academy of Sciences, which Chambers cites. Had Chambers desired to illustrate how unscientific scientific societies can be, he could hardly have done it in a more striking way. Here are the cases.

1. A woman was cured by bleeding her, tying her to a chair for a year and feeding her bread and water.

2. One patient was tied to a tree and two hundred pails of water thrown on him.

3. A girl was given a bath in river water with a bushel of salt thrown in. They plunged her naked again and again until she was senseless, and then left her there. When she

[53] According to another source, "The cock was to be seized while crowing, the tip of the spine to be cut off, and the rear part of his body to be applied at once to the imposthume, simultaneously, if possible, with his being killed by suffocation." Campbell, A. M., *The Black Death and Men of Learning*, New York, 1931, p. 89.

came to, she looked at the water without any concern, and apparently was cured.[54]

After such a description of the treatment of hydrophobia it is useless to speak of any scepticism concerning materia medica in Chambers. All the treacles and theriacs and electuaries are accepted without question. Unlike Venel in *L'Encyclopédie,* he makes no attempt to criticize the vastness of the number of drugs and plants that go to make up these confections.[55] Not only that, but he argues very strongly in favor of the therapeutic value of gems which are included in so many of those concoctions.[56] Thus the treatment of medicine by Chambers seems the weakest from the point of view of modern science, and is still very close to ancient and medieval medicine. But medicine after all was perhaps the last of the arts to assume a scientific attitude, and at the close of the seventeenth century, despite the great progress made in the subject, medical men were still groping in the dark in many branches of it.

There remains to note the survival in Chambers of the belief in occult science, magic and allied arts. A good deal of scepticism is noticeable but not without an admixture of credulity. Thus in the article, " Occult ", the statement is made that " weak philosophers, when unable to find the cause and effect or unwilling to own their ignorance, say it arises from the occult virtue, an occult cause or an occult quality." But how can we explain (this in another section) the antipathy existing between the salamander and the tortoise, the vine and the elm, the sheep and the wolf, the olive and the oak? Or why do hens fly at the sound of the harp, strung with fox-gut strings? The answer is that there are certain

---

[54] See article, " Hydrophobia ".

[55] See article, " Confection of hyacinth ", " electuary ", " hiera picra ", " mithridate ", " treacle ", " theriacs ", and the like.

[56] See article, " Ferns ".

occult qualities inherent in bodies which create antipathies.[57] One is tempted to ask whether Chambers intended to include himself among those weak philosophers.

Similarly Chambers discredits the efficacy of charms, philtres and amulets. But the evidence in favor of amulets seems to be so strong that it nullifies his own scepticism in the matter. An amulet, says Chambers, is "a kind of external medicament, wore about the neck or other part of the body, to prevent or remove diseases. Such are quills of quicksilver, or arsenic, which some hang on the neck or wear under the shirt against the plague and other contagious diseases; and that wore by women of the East Indies to bring down the menses." Although he asserts they have lost their old reputation, he cites Robert Boyle's favorable testimony. That scientist found the moss of a dead man's skull, though only applied so as to touch the skin, the most effective cure in checking nosebleed. "The great Robert Boyle", he says, " alledges them [i. e. amulets] as an instance of the ingress of external effluvia into the habit, in order to show the great porosity of the human body." In other words, according to Chambers, Boyle seems to have found a scientific explanation for an old superstition.[58]

Also the evidence of a chief physician in Moravia tends to prove that amulets are after all very effective. This physician found that troches of toads worn as amulets not only preserved him and his friends from the plague, but when they were put on the plague-sores of the others, greatly relieved and even effected a cure. Moreover Boyle says that the effluvia of even cold amulets in time pervades the pores of the living animal, thus effecting a cure.[59] As for philtres, Chambers waives aside Van Helmont's testimony in their

[57] See article, "Salamander".
[58] See article, "Amulet".
[59] See article, "Amulet".

favor, characterizing it as mere cant and asserting that all philtres are chimeras.[60]

In the article on magic the cyclopedist does not seem to object so much to natural magic as to celestial and goetic or superstitious magic. Natural magic he claims is no more than the application of natural active causes to passive subjects, and in this manner many surprising but yet natural effects are produced. On the other hand celestial magic borders on judicial astrology and that is condemned. Goetic magic consists in the invocation of devils, the effects of which are evil and wicked. "The truth is, however", says Chambers, "these have not all the power that is given or that is usually imagined; nor do they produce half those effects ordinarily ascribed to them." Although divination is not condemned as outrightly as in the *Encyclopédie* by Diderot, yet there is no question as to Chambers' discrediting that art.[61] After classifying divination as artificial and natural, he points out that natural divination is founded on the superstition that the soul has from its own nature some foreknowledge of the future. The Church Councils are cited as having condemned divination, since they supposed those who pretended a knowledge of the art to be in league with the devil.

[60] See article, "Philtre". Chambers distinguishes between the true and the spurious, but in the end he repudiates both kinds. He says: "The true philtres are those supposed to work their effect by some natural and magnetic power. There are many grave authors who believe the reality of these philtres... Van Helmont says that upon holding a certain herb in his hand, and taking afterwards a little dog by the foot with the same hand, the dog followed him wherever he went and quite deserted his other master." Helmont's explanation is that the heat communicated to the herb, animated by emanations of the natural spirits, "determines the herb toward the man and identifies it to him; having then received this ferment it attracts the spirit of the other object magnetically and gives it an amorous motion." "But this," adds Chambers, "is mere cant, and all philtres whatever facts may be alledged are mere chimeras."

[61] See article, "Divination".

Although he characterizes chiromancy (a form of divination) [62] as a vain and trifling art, he does not deny completely that a person's inclinations may not be revealed in his hand; for there is, it seems, a near correspondence between the parts of the hand and the internal parts of the body such as the heart and the liver where the poisons reside. The art of necromancy is defined without any comment. Since divination in general is discredited by Chambers, we might assume that that form of divination too is considered valueless.

And yet sceptical as Chambers is of divination, he does not hesitate to accept withcraft and fascination. " There may be some foundation for what we call fascination and witchcraft. We have infinite instances and histories to this purpose, which it were not fair to set aside simply because they do not reconcile with our philosophy. But as it happens there seems to be something in our philosophy to countenance them ". His explanation is based on the theory that all animals emit effluvia through the pores of the skin and by the breath, or through other parts. The eye too emits effluvia. This emission from the eye strengthened by the subtle juice from the optic nerve, becomes a fluid of volatile matter that can fly instantaneously through solid capillaries and have its effect wherever it strikes, that effect being limited by distance, the impetus of the eye, the quality of the juices and the delicacy or coarseness of the object it falls on. In this fantastic way Chambers desires to account for the phenomena of withcraft and fascination.[63]

Physiognomy [64] is not a pretended art. It is the art of knowing the humor or disposition of a person from observation of the lines of the face. " Of all the fanciful arts of the ancients discussed among the moderns, there is none that has

[62] See article, " Chiromancy ".
[63] See articles, " Witchcraft " and " Fascination ".
[64] See article, " Physiognomy ".

so much foundation in nature as this." The explanation in brief is that the animal spirits, whenever affected by some object, transmit the affection to the brain whence it is carried by the nerves to that part of the body " as is most suitable to the design of nature." The result of this change is noted in the face, which acts as a sort of dial plate.

CHAPTER III

ZEDLER'S *Grosses Vollständiges Universal Lexicon* AND THE HISTORY OF SCIENCE

BEFORE proceeding directly with the analysis of Zedler's *Grosses Vollständiges Universal Lexicon,* a few general considerations concerning the work are in order. In the first place, unlike Ephraim Chambers, who had a definite interest in science,[1] and Diderot and D'Alembert who made contributions of their own in the field of science,[2] the editors of the *Lexicon* were not, so far as we know, men of science. Johann Peter von Ludewig, for example, was interested in constitutional law;[3] P. D. Longolius [4] another collaborator, was chiefly interested in philology, education and history. Hence, although the sciences are included, it is doubtful whether science was uppermost in the minds of the editors. Considering the number of volumes (64) they published, one would conclude that their aim was all-inclusiveness rather than scientific specialization. And in that respect their work was unique and an innovation for its age. Nevertheless, this very desire for all-inclusiveness compelled them to give much space to the sciences and related fields.

[1] See Chapter II.

[2] See Thorndike, "L'Encyclopédie and the history of science," *Isis,* VI (1924), pp. 361-86.

[3] See Ludewig (1668-1743) in *Allgemeine Deutsche Biographie*; also Zedler's *Lexicon,* vol. xviii.

[4] See Longolius in Brockhaus, *Konversations Lexicon*: according to Brockhaus, Longolius was interested in philology, education and history. He mentions also Frankenstein. Longolius (1704-1779) in the *Allgemeine Deutsche Biographie* is described as philologist, educator, historian and also a collaborator of the *Lexicon.* Brunet, J. C., *Manuel de Libraire* (Paris, 1860-65), vol. vi, p. 1847, no. 31865 also mentions these three only.

Another point, already mentioned above, is the cooperative nature of their work. Throughout the middle ages and early modern times, encyclopedias were always the result of individual effort. In 1732 for the first time collaboration is found essential. Even at that time the idea seemed so unusual that the editor (P. von Ludewig) took special pains to explain the advantages of a cooperative work.

It is true [says the editor] that a learned man ought to know everything, because all sciences (*Wissenschaften*) are related to one another, and one subject is always stretching out a hand to the other. But it is one thing to have a general understanding of all sciences, quite another thing to be master in all fields, and still another to have sufficient time for everything.[5]

To facilitate therefore the production of the work nine specialists were chosen, the number to accord with the number of the muses.[6] (Here again is an interesting sidelight on the non-scientific interest of the editor). For the purpose of this study, a knowledge of the names of those engaged to write the articles on the various sciences and related fields would be of a considerable advantage. But of the nine specialists, the three that could be identified, namely P. Ludewig, Longolius and Frankenstein, were all interested in the arts,[7] so that it is difficult to determine the degree of expertness employed in the compilation of the scientific material in the *Lexicon*.

A third point to be noted is the theological bias, which

[5] See Foreword (*Vorrede*, p. 6): "Ein gelehrter Mann muss zwar alles wissen, weil alle Wissenschaften mit einander verschwestert, u. eine immer der andern die Hand biethet; nur ein anders ist es einen Begriff von allem zuhaben wieder ein anders in allem Meister zu sein, u. wieder ein anders hinlängliche Zeit zu haben alles allein abzuwarten."

[6] See *Vorrede*, p. 6: "Er hat nach Anzahl der IX Musen neunerley gelehrte Leute... gedinget."

[7] Brockhaus, *Konversations Lexicon*, art. "*Enzyklopedie*".

again marks it off not only from Diderot's *Encyclopédie* but from Chambers' work too. Chambers was a freethinker and Diderot a deist. Although his critical attitude toward Christianity did not prevent Diderot from accepting the Old Testament chronology, nevertheless it has been pointed out that " new matter (mechanical arts etc.) . . . crowded out or reduced the relative importance of theological interests." [8] The editors of the *Lexicon* not only allot generous space to matters theological, but betray an orthodox point of view which obscures their vision on many an occasion. It is evident that these editors were not at all in sympathy with the free-thinking spirit that pervaded certain enlightened circles in Germany at that time. To cite a few instances. There were free spirits in eighteenth century Germany who denied the existence of the devil, who identified the soul and the brain as one, who questioned the truth of revelation and denied the possibility of miracles.[9] The *Lexicon* in all these matters adheres to the usual orthodox viewpoint. Hence as a repository of its civilization it would seem to tell us that the prevailing culture had not been much changed or stirred by the winds of free doctrine.

And yet it is unfair to say that the *Lexicon* is forever looking backward. The new science is accepted. But the old theology and old beliefs of all sorts are not entirely rejected. This is noted in definitions and classifications. In the article on history [10] that subject is divided into church history, political history, natural history, history of the arts and miscellaneous history, whereas in the article on the sciences [10a] it is divided into sacred and profane

[8] Thorndike, L., " L'Encyclopédie and history of science," *Isis*, VI (1924), p. 374.

[9] Biedermann, K., *Deutschland im Achtzehnten J. H.* (Leipzig, 1880), vol. ii, part i, pp. 397-399.

[10] Vol. xiii, article, "*Historie*", p. 281.

[10a] In article, "*Wissenschaften*", vol. 57, p. 1401.

history. Something has been gained from the new learning, but the old division can be detected in the new. On the other hand the providential history of the medieval period is gone. The editors define history as experience; and say that " he who recognizes the present as having its origin in the events of the past, will readily recognize the usefulness of history ".[11] A good historian furthermore " must be wise, experienced, critical in the reading of his sources and objective ". The difficulties of the historian are well appreciated by the editor when he states that it is impossible to be a good historian, for in order to be one, " one must belong to no order, to no party, no country and no church ".[12] Nevertheless, the historical perspective in the *Lexicon* is marred, first by an acceptance of the Old Testament as an accurate historical source, and secondly by the prevailing eighteenth century attitude toward the middle ages. The middle ages were the dark ages. Only a few great individuals relieved the monotony of darkness with their candle light of knowledge. Thus the happy generalization that " the present has its origin in the events of the past " is not understood fully enough to allow for the notion of continuity or for the critical examination of the prevailing opinion with reference to the immediate or remote past. To cite one instance at this point, natural philosophy in the middle ages is uncritically summed up with the statement that the scholastics accomplished nothing in the field and that Roger Bacon and Albertus Magnus alone sought to discover fundamental principles by means of chemical experiments.[13]

[11] Vol. xiii, p. 282, article, "*Historie*": " Wer erweget dasz die gegenwartigen Dinge in dem Vorhergehenden ihren Grund haben... der selbe wird an die Nutzen der Historie nicht zweifeln."

[12] See article, "*Historie*", vol. xiii, p. 286.

[13] See article, "*Natur-Lehre*", vol. xxiii, p. 1162. In speaking of natural philosophy in the middle ages, it says: " Doch fanden sich einige bey dieser so grossen Finsternisz welche wieder ein Licht aufzustecken

The aforementioned attitude toward the middle ages is even more clearly brought to light in the treatment of the history of astronomy.[14] It is indeed full of the usual errors and lacunae. Pythagoras is again presented as the first among the Greeks to suggest the heliocentric theory. Although the Arabs are credited with having assimilated Greek astronomy, their own work in the field is omitted, as for example, the astronomical tables of Al-Kwarizmi or the early work of the Arab, Mashallah (Messahalla). Nor for that matter is there any definite statement that the *Almagest* had been translated from the Arabic in 1175. It so happened that although a translation had been made from the Greek somewhat earlier (1160), it was the translation from the Arabic that was used by the Europeans. The fact that Europe learned its astronomy from the Arabs may not in itself be very important, but coupled with all the other facts relating to the transmission of Greek science to western Europe, it does aid in placing the medieval and Arabic civilizations in their proper perspective. As for European contributions during the middle ages, the tendency is to prefer the renaissance mathematicians and astronomers to their predecessors. Thus George Peurbach (1423-1467) and Regiomontanus (1436-1476) are given some attention. But as for their translation of the *Almagest* from the Greek, their aim to be sure was to eliminate the medieval barbarisms of the older translations of the thirteenth century. Yet their version " was not a complete and exact translation of the Greek but merely an epitome of it ".[15]

This undue renaissance bias blinds the editors to any

bemüht waren, in dem sie durch chymische Versuche die selbständige Principien zu entdecken suchten, wodurch die Physick in eine neue Veränderung kam."

[14] See vol. ii, article, "*Astronomia*", p. 1963 *et seq.*

[15] Thorndike, L., *Science and Thought in the Fifteenth Century* (New York, 1929), p. 145.

earlier substantial contributions to astronomy. Thus they seem to be unaware of the fact that the Ptolemaic astronomy had been questioned in the fourteenth century by a few scholars in the University of Paris. Nicholas Oresme (1377) for one, thought that " one could well sustain [the view] that the earth is moving in a diurnal movement and not the heavens ".[16] In the fifteenth century Nicholas of Cusa made a similar suggestion. Other less conspicuous names might have been mentioned in such an all inclusive work. For the early Middle Ages Martianus Capella's book on the seven arts might have been included. His modified heliocentric theory was adopted by Tycho Brahe. Although mention is made of the *Alfonsine Tables* of the thirteenth century, the astronomical tables of Abraham Zacuto produced in the fifteenth century in the University of Salamanca are omitted.

So much for the pre-Copernican period. The Copernican or modern period is, it is true, treated more fully and better. Tycho Brahe is somewhat overpraised as one who far excelled all previous astronomers. We are told of his vast supply of astronomical instruments in his almost royal castle at Uranienberg. Yet his opposition to Copernicus is not mentioned. It is perhaps interesting to note Tycho's argument against the heliocentric hypothesis; an argument as old as the Ptolemaic system itself. In the first place, he said, the swift motion of the earth on its axis would shatter it, and secondly that as the earth moved from one end of the orbit to the next a change in position of the fixed stars should be apparent. This contention could not be very easily dismissed by the adherents of the heliocentric theory. It was not until 1838 that parallax of the fixed stars was

[16] Duhem, P., "*Un precurseur français de Copernic, N. Oresme*," in *Revue générale des sciences pures et appliquees*, Paris, 1909, Année 20, no. 21, pp. 866-873.

proved. It was perhaps for this reason that Tycho's modified explanation was more widely held in his own time than that of Copernicus. Tycho's contribution to a proper knowledge of comets is recounted in detail in the article, 'Comet'.[17] He is given full credit for showing that the comet's orbit lay far beyond the moon,[18] and "that the diameter of the earth has no noticeable relation to the comet's distance from the earth."[18a] This leads the lexicographer to deny the oft-accepted belief that comets are presages of misfortune since " they are bodies that have a regular motion like the rest and can only appear to us on certain occasions, when in the course of their travels they come near us. Why should this time be more unfortunate than that during which we see no comet?"[18b]

As for Kepler, the lexicographer does state that from Tycho's many observations Kepler inferred " that the planets did not move in eccentric but elliptical orbits in which the sun occupied one focus."[19] Without any further explanation, the lexicographer concludes that this important inference strengthened the Copernican theory. It would seem that the full implication of Kepler's Laws of elliptical orbits is not brought out clearly enough. The fact is that the Copernican system was in some respects Ptolemaic in character,

[17] See article, "Comet", vol. vi.

[18] *Ibid.*, pp. 799-800.

[18a] *Ibid.*, p. 800: "dasz der Diameter der Erde gegen die Weite des Cometens von der selben keine merkliche Verhaltnisz habe."

[18b] *Ibid.*, vol. vi, p. 812: "Sie sind Welt-Körper die eine regulaire Bewegung wie die andern haben, und unseren Augen nur zu gewissen Zeiten erscheinen können, wenn sie nemlich in ihrer Bewegung den selbigen naher kommen. Warum soll diese Zeit unglücklicher seyn als jene, da wir einen Cometen nicht sehen!"

[19] See "*Astronomia*", vol. ii, p. 1966: "Es zeigte der selbige (i. e. Kepler) aus denen Observationibus des Tychonis, dasz die Planeten sich nicht in denen Eccentrischen Circeln sondern in Ellipstischen Bahnen in deren einem Foco die Sonne stehet, bewegten."

for, in accepting the circular motion of the planets, it could not "save the phenomena" unless it retained Ptolemy's eccentric and epicyclical motions, though somewhat reduced in number. In other words Copernicus considered the whole astronomical problem as purely kinematic, though he thought that he explained how the planets actually moved. But Kepler's system, though still kinematic, pictured motions that were later proved mechanically explainable; his laws being fully consistent with the mechanics of circular and elliptical motion, established by Newton.

The subject of astrology undergoes much the same treatment it received in Chambers' and Diderot's works. There is the usual definition that

astrology is an art according to which one seeks to derive both the natural effects of the weather, the production of plants from the earth, as well as influences and union (conjunction) of the constellations, in order to demonstrate through them the necessity of past occurrences as well as to foresee future events." [20]

After an extended excursus into the history of the art, the lexicographer divides it into natural and judicial astrology, accepting the former, and listing objections to the latter. It may be true, he argues, that the sun and moon bring about a change in air and in this manner produce a varied effect on earthly things; and it is true that the sun and moon have an influence on disease. But such suppositions by no means validate the whole art. Besides, ethical and religious grounds for opposition to the art are far more cogent than

---

[20] "*Astrologia*", vol. ii, p. 1953: "Astrologia ist eine Kunst nach welcher man so wol die natürlichen Würckungen des Wetter u. Hervorbringung der Gewächse aus der Erde als auch die Handlungen u. Zusammenfügung der Gestirne herleiten will, um so wol die Notwendigkeit der Geschehenen Sachen daraus zu beweisen als Zukünftige vorherzusehen."

any naturalistic argument, for to accept the validity of judicial astrology is equivalent to denying free will and God. One might well ask if such reasoning differs essentially from that of the scholastics that were opposed to the art. Another strong argument is from authority. Pico della Mirandola [20a] who wrote against the astrologers argued that the coincidental relationship between stars and human fate was only an accident. Paracelsus [20b] furthermore alleged that astrologers had a compact with the devil. Bayle [20c] is quoted as having made a similar allegation. But then the lexicographer continues, it is not necessary to go so far as to attribute to the practitioner of the art connivance with the devil, although it cannot be denied that those persons devoted to astrology also devote themselves to the other occult sciences. But there are more cogent arguments against the art. On closer examination one finds that it is only idle curiosity to seek knowledge of the past which no longer affects us (here is a flat contradiction of the attitude expressed with reference to history). And to know the future, the events of which are inevitable, reason forbids, for did not Favorinus say that in our ignorance of the future is revealed the distinction between men and gods? In other words all the arguments advanced against astrology do not explain the material baselessness of the entire presumption; which fact should have been plain to all after the astronomical discoveries of Kepler and Galileo. Instead, the opposition still rests on the old religious and ethical grounds.

Furthermore, the fact that natural astrology is accepted strengthens the conclusion that the full implications of the astronomical works of Kepler, Copernicus and Galileo were

[20a] *Ibid.*, p. 1961: " Joh. Picus Mirandulanus will die Zutreffung eines Exemple dem Zufall zuschreiben."
[20b] *Ibid.*, p. 1961.
[20c] *Ibid.*, p. 1961.

not entirely understood even in the eighteenth century. "In what pertains to natural things," we read in the *Lexicon*, "the influence of the planets must be conceded."[21] The explanation given is that the rays of the planets influence the earth. This takes place through a reflection of the rays (*Zurückprallung*) and therefore *per accidens*. Although it is conceded that it does not seem probable, nevertheless it is not considered impossible.[22] Such reasoning does not differ much from the naturalistic argument justifying the influence of the planets on human life. In the sixteenth century it was claimed that the planets in their revolutions emit vapors that penetrate the animal spirits of men, and stir the passions to uncommon heat, with the result that wars and revolutions follow. It was for this reason, perhaps, that Kepler believed in astrology,[22a] and cast horoscopes for a living.

The treatment of natural philosophy suffers from the same historical misconceptions that occur elsewhere in the *Lexicon*. It is said that this study had its origin among the

[21] *Ibid.*, p. 1959: In section on natural astrology: "Was die natürlichen Dinge anbelangt, so kan man noch einen Einflusz dern Planeten zu geben". The lexicographer refers the reader in the same article to Ridiger: *Physica Divina*, who explains the possibility of such an influence in that " die Strahlen der selben unsere Erde beruhren konnten ... er (Ridiger) sagte aber dasz dieses nur durch die Zuruckprallung u. also nur per accidens geschehe, daher es denn nicht wahrscheinlich ware, dasz diese Strahlen per accidens eine solche Wirkung haben konnten, unerachtet es doch dasselbe vor keine Unmoglichkeit ware."

[22] At the same time in the article, "*Astro-Meteorologia*", vol. ii, p. 1963, it is denied that the aspects of the planets are in any way related to weather conditions on earth.

[22a] Burtt, E. A., *Metaphysical Foundations of Modern Physical Science* (New York, 1925), p. 58, says that those who maintain that Kepler did not thoroughly believe in astrology "can hardly have read his essay *De Fundamentis Astrologiae Certerioribus* in which he advances for the criticism of philosophers, seventy-five propositions, of varying generality, whose soundness he is prepared to defend."

Hebrews.[23] Neither the patriarchs nor Moses were physicists, but Solomon was, and whatever he wrote of nature he wrote with a divinely endowed wisdom with which philosophical erudition can not be compared.[23a] The fact that we have no record of Solomon's marvelous achievements in science does not seem to puzzle the editors at all. When the Ionian philosophers, on the other hand, sought to understand natural phenomena by the light of reason, their physical systems invariably led to atheism, and " the true attributes of an atheist are unreason, and folly ".[24] Aristotle's physics again, is found wanting, because it was founded on metaphysical speculations and not on " real principles ".[25] Ever since the time of Francis Bacon, the champion of inductive reason, the idea had been widely held that the greatest shortcoming in Greek science was the neglect of experimentation. The Greeks simply neglected experience and just theorized, so it was said. There is ample evidence to expose the absurdity of such a notion, and to show that Greek philosophers from the Ionians to Aristotle, as well as those of the Alexandrian school, derived their general principles from observations and experimentation. So far as Aristotle is concerned one need only recall any number of his treatises which are largely collections of facts such as, for example, his *History of Animals*. One need but mention his experiment on the development of the chick to indicate that he did not spin his theories from thin air. In his treatise on

---

[23] The history of science is treated in the article " *Natur-Lehre*," vol. xxiii, pp. 1156 *et seq*.

[23a] It is interesting to note that Roger Bacon too believed that complete knowledge and understanding were vouchsafed by God to Solomon.

[24] See article, " *Natur-Lehre* ", vol. xxiii, p. 1157: " Die eigentliche Eigenschaften eines Atheisten sind unvernunft u. Thorheit."

[25] " *Natur-Lehre* ", vol. xxiii, pp. 1158-1159: " Das haupt Versehen war, dasz er metaphysische u. keine reale Principien zum Grund geleget."

logic he makes his position quite clear. In speaking of the rules of reasoning, he says:

The way must be the same with philosophy as it is with respect to any art or science whatever; we must collect the facts and the things to which the facts happen in each subject and provide as large a supply of these as possible.

Again,

we arrive at knowledge either by induction or demonstration. Demonstration proceeds from universal theoretical propositions, induction from particulars. But we cannot have universal theoretical propositions except from induction.[26]

Hence both are essential for the proper pursuit of science, the idea and the facts. Neither one is conceivable without the other. The point is, therefore, that Aristotle's scientific generalizations were not grounded on purely metaphysical speculations. Both his theory and his practical achievements in science refute the notion expressed in the *Lexicon*. If Aristotle's physics erred, it was not because it lacked a practical basis, but more properly because "the ideas were not distinct and appropriate to the facts."[27]

An analysis of the *Lexicon's* treatment of natural philoso-

[26] Whewell, *History of the Inductive Sciences* (London, 1837), vol. i, p. 74.

[27] *Ibid.*, p. 80. The author gives the following instance of Aristotle's method. Aristotle tries to explain why when the ray of light passes through a hole, the image formed is always round, no matter what the shape of the hole. Says Whewell: "We shall easily perceive this appearance to be a necessary consequence of the circular figure of the sun, if we conceive light to be diffused from the luminary by means of straight rays proceeding from every point. But instead of this approximate idea of rays, Aristotle attempts to explain the fact by saying that the sun's rays have a circular nature which it always tends to manifest; and this vague and loose conception of a circular quality employed instead of the distinct conception of rays which is really applicable, prevented Aristotle from giving a true account of even this very simple optical phenomenon."

phy in the middle ages necessarily results in a criticism of its uncritical attitude toward the period, and the consequent omissions. Such a procedure may not be altogether fair. But the persistence even in our day of the Renaissance attitude toward the medieval period, which a competent historian like John Bury still termed "a thousand years of darkness" [28] justifies such procedure even if it does entail a repetitive account of the result of critical historical research. Briefly then, the *Lexicon* states that the scholastic philosophers were much too preoccupied with philosophy to attempt anything useful (practical) in the sciences; but there were a few (Roger Bacon and Albertus Magnus) in this dark age " who made an effort to rekindle the light in that they sought to discover independent principles by means of chemical experiments; in this fashion science took on a new form." [29] In this brief statement the customary inaccuracies have crept in. In the first place the mistake is made of considering men like Roger Bacon and Albertus Magnus as anomalies of their age rather than as outstanding representatives; secondly, the error of overestimating their efforts and achievements; and thirdly the complete ignorance of the work of others in the field. Thus Roger Bacon can be better understood as one of a group than as a unique figure. In the *Opus Minus* he says to the Pope: " I confess that there are several men who can present to your wisdom in a better way than I can these very subjects of which I treat." [30] In speaking of mathematics in the study of science, again Bacon lets us know that " there were found some famous men as Robert (Grossteste), Bishop of Lincoln, Adam Marsh and

[28] Bury, J., *History of the Freedom of Thought* (London, 1913), p. 52. He characterizes the middle ages as the " millenium in which reason was enchained, thought was enslaved, and knowledge made no progress."

[29] See above note 13.

[30] Thorndike, L., *History of Magic and Experimental Science* (New York, 1923), vol. ii, p. 643.

some others who knew how by the power of mathematics to unfold the cause of all things, and to give a sufficient explanation of human and divine phenomena." [81] And there were others whom Bacon did not mention, as for example John Peckham and Bartholomew of England.[82] Then again it is questionable to what extent science took on a new form as a result of "experimental" work of Roger Bacon and Albertus Magnus. It is for instance doubtful if Bacon's understanding of "experimental science" was in any sense equivalent to our own. Says Thorndike:

> he seems to regard it (i. e. experimental science) as something distinct from the other natural sciences, such as optics, alchemy, astronomy, rather than as an inductive method through regulated and purposive observation and experience to the discovery of truth, which should underlie and form an essential part of them all. . . . It is not, like modern experimentation, the source but 'the goal of all speculation.' It is not so much an inductive method of discovering scientific truth as it is an applied science, the putting the results of the 'speculative' natural sciences to the test of practical utility.[83]

And as for his work in optics, even as an applied scientist, it is well known now that burning glasses and spectacles were common in his time, and that he had little to do with these inventions. Also there is no evidence that he made a telescope.

Similarly although Albertus Magnus and his associates performed experiments with animals, it is doubtful if he performed any well planned experiments in physics. In his *History of Botany*, Meyer says of him: " I do not know of his undertaking an experiment in order to solve a physiological or physical problem in which he had a clearly defined

---

[81] Singer, C., *From Magic to Science* (New York, 1928), p. 91.

[82] Thorndike, L., *op. cit.*, vol. ii, p. 629.

[83] *Ibid.*, vol. ii, p. 650.

purpose and the suitable materials at hand for carrying it out; his books on plants certainly do not contain a single one."[34] It is doubtful, therefore, if the so-called experiments of such men as Albert or Bacon produced any fundamental change in the sciences, as the *Lexicon* would have us believe. On the other hand we need not conclude that modern physics began in 1638 when Galileo published his *Two New Sciences,* as one modern writer would have it. We know, as a result of the researches of Duhem, that Leonardo da Vinci, who understood the principle of the inclined plane, the principle of moments, the principle of equilibrium and the principle of virtual velocities, had his precursors in the fourteenth century. The Nominalist school in that century not only cultivated the experimental sciences but took a critical attitude toward Aristotelian physics. William of Occam, for example, criticized Aristotle's explanation of the motion of a projectile. Then toward the middle of the century, Jean Buridan posited the principle of impetus, which applied equally to both celestial and terrestrial motion. In this manner, we may say, he laid the foundation for the later development of dynamics.[35]

In discussing the atomic theory, the *Lexicon* proceeds forthwith to an account of the seventeenth century revival by Boyle, Gassendi, and Du Hamel who accepted atoms as fundamental elements of matter, but rejected the philosophical materialism of Epicurus. The criticism here is the implication that the atomic theory died with Lucretius and that it was not revived till the end of the seventeenth century. And yet even the Arabic scholastics adopted a modified version of the atomic theory, since they introduced a creator who

---

[34] *Ibid.,* vol. ii, p. 539.

[35] Duhem, P., *Etudes sur Léonard de Vinci* (Paris, 1913), vol. iii, pp. vii-ix.

[35a] "*Natur-Lehre*", vol. xxiii, pp. 1149 *et seq.*

created and destroyed the atoms. They even acknowledged the existence of voids between atoms, although some Mutakallimun insisted on the continuity of matter.[36] In twelfth century Europe there were two Christians who seemed to have adopted atomism. Adelard of Bath hinted at it in one of his books.[37] And William of Conches in his *Elements of Philosophy* asserted that all bodies are composed of elements; and an element he defined as the smallest particle of any body, simple in quality and small in quantity, which can never be seen and can only be understood as a concept of division.[38] It is interesting to note that Boyle's definition

[36] Lasswitz, K., *Geschichte der Atomistik* (Hamburg, 1890), vol. i, p. 137.

[37] In Adelard's book *De Eodem et Diverso*, we find this passage: "Whose eye can embrace the infinite space of heaven? Whose ear can hear the harmonies? What eye can distinguish the atoms? What ear can distinguish the noise of their collision?" Lasswitz, *op. cit.*, p. 71 cites this passage from the French translation of A. Jourdain: "*Recherches Critiques sur L'âge et L'origine des Traductions Latines*" (Paris, 1843), pp. 68-69.

The original Latin reads as follows: Quis enim unquam caeli spatium visu comprehendit? Quis sonum eiusdem caelestemque concentum auribus clausit? Quis item atomi parvitatem oculo distinxit? Quis sonum eisdem atomis collisis creatum aure notavit?, p. 13, *De eodem et Diverso* (first printed in 1903 in *Beiträge zur Geschichte der Philosophie des Mittelalters*, Band IV, Heft 1).

[38] Lasswitz, *op. cit.*, pp. 74-75, cites the following passage from William of Conches: *Elementis Philosophiae*, p. 209. "Omnia corpora ex elementis constant ... Elementum vero definiunt philosophi, est simpla et minima alicujus corporis particula; simpla ad qualitatem minima ad quantitatem ... Haec elementa nunquam videntur, sed ratione divisionis intelliguntur."

Burtt in his *Metaphysical Foundations of Modern Physics*, p. 41, notes that William of Conches stressed the geometrical atomism of Plato in the *Timaeus*. Nevertheless even this atomism was opposed by the theologians. Walter of St. Victor writing about 1180 against Peter Abelard and three other heretics refers to William of Conches as having adopted the atomic theory and adds: "We condemn and excommunicate their atoms, the rules of the philosophers and all other ridiculous theories of this sort."

was not essentially different from the foregoing twelfth century conception. In his *Sceptical Chymist*, Boyle says: " I mean by elements certain primitive and simple bodies or perfectly unmingled bodies; which not being made of any other bodies or of one another are the ingredients of which all those perfectly mixed bodies are immediately compounded and into which they are ultimately resolved." [39] Such evidence would seem to substantiate the view that despite the attack on Aristotle's four elements in the seventeenth century, there was no fundamental change in the atomic theory till Dalton. Furthermore from the twelfth to the eighteenth century, philosophers and scientists were divided in their opinions, some accepting, others rejecting atomism. Thus Roger Bacon criticized the theory adversely. On the other hand Nicholas of Cusa, in the fifteenth, and Fracastoro, in the sixteenth century, accepted it.[40] And as late as the eighteenth century it was still argued by some physicists that the " fact " of God's omnipresence precluded the possibility of a void in nature.[41]

But let us turn from the discussion of theory to the question of aims and objectives in the study of science. Inasmuch as the primary aim is to combat successfully atheism and superstition, the true physics, we learn, serves the important purpose of bringing about a conscious recognition by man of the divine presence and the divine nature of things.[42]

[39] Boyle, R., *The Sceptical Chymist*, in *Works* (London, 1744), vol. i, p. 350.

[40] Smith, Preserved, *A History of Modern Culture* (New York, 1929), p. 87.

[41] See article, "*Raum*" (*leerer*) "*Vacuum*", vol. xxx, p. 1120, the encyclopedist cites Rudiger: "*Physica Divina*," saying as follows: " Rudiger erinnert dasz man sich keinen solchen leeren Raum einbilden könne, dasz darinnen gar keine Substanz anzutreffen, indem solches wider die Allgegenwart Gottes wäre."

[42] See article, "*Natur-Lehre*", vol. xxiii, p. 1153: " Die wahre Physick

Aside from this aim, science has its practical usefulness in medicine, because it is helpful in the preservation of health; in ethics, wherein sound reason must recognize the will of God through nature; and in logic wherein abstractions must find an adequate basis in concreteness.[42a] The most obvious of all aims, namely the effort to discover the laws of nature through disinterested experimentation, is omitted. One might well ask what has been gained by the work of the seventeenth century scientists? It would seem that the scientific achievements and attitude of such men as Galileo, Boyle and Newton did not prevent religious bias and medieval empiricism from continuing their sway over men's minds even in the middle of the eighteenth century.

In his discussion of method in science the lexicographer asserts: " in the study of natural phenomena one begins with the particular, that is, one first recognizes the particular natural phenomena and their effects, then one arrives at the causes, and finally at general conceptions."[43] It is tempting to take issue with such a generalization, for none of the great scientists, had they proceeded along such purely inductive lines, would have gotten very far in establishing or discovering any natural laws. Kepler, Newton and Harvey, to mention three outstanding figures, all formulated hypotheses first and then had them tested. Thus Kepler proceeded on the hypothesis that all natural laws could be expressed in exact numerical terms.[44] Similarly Harvey

dient zu einer Erkänntnisz Gottes und Göttliche Dinge vomit man dem Aberglauben so wol als der Atheisterey begegnen kan."

[42a] See article, "*Natur-Lehre*", vol. xxiii, p. 1154.

[43] See article, "*Natur-Lehre*", vol. xxiii, p. 1155. "In der Erkänntnisz der natürlichen Dinge fangt man von Besondern an, das ist man erkannt die besonderen natürlichen Phoenomena u. Wirkungen u. kommt auf die Ursachen, folglich auf die allgemeinen Anmerckungen."

[44] Bancroft, W. D., *Methods of Research*, p. 178 (Rice Institute pamphlet, vol. xv, Oct., 1928, pp. 167-286).

began with his hypothesis of the circulation of the blood long before he had all the data that verified it.[45] And Newton said that "no great discovery was ever made without a bold guess".[46] The point to be particularly noted here is that the *Lexicon's* generalization is largely the result of Francis Bacon's overemphasis of the possibilities of induction, and the complete neglect of the deductive process involved in the scientific method. For the mere collection of data or facts will never reveal the generalization, since that does not lie, as it were, imbedded in the facts, waiting for the fact gatherer to discover it; rather it is something with which the scientist approaches the data at his disposal, or which he conceives in attempting their classification.

The *Lexicon* falls into the same error in its discussion of the importance of experiment.[47] In physics, it is claimed, it is impossible to arrive at valid propositions concerning causes before making a careful study of the results of experimentation. It is obvious, to be sure, that speculative hypotheses that are not verifiable through experimentation can be of little value. What the lexicographer fails to see is that experimentation without an hypothesis has no more meaning than an unverified hypothesis.

Shortcomings such as the foregoing anent purpose and method in science do not prevent the editors from displaying sound judgment and a scientific attitude in their discussion of different branches of physics such as light, optics and magnetism, although here too the historical sections are found wanting.

In their treatment of the corpuscular and undulatory theories of light, the editors find it hard to choose between them, since both approximate an explanation of experience, and both have their difficulties. But they lay down a sound

---

[45] *Ibid.*, p. 177.   [46] *Ibid.*, p. 179.
[47] See article, "*Experimentum*", vol. viii, pp. 2344-45.

basis for choice when they suggest that in choosing a physical hypothesis one must select the one which explains most of the phenomena. Even though the selected hypothesis has its shortcomings, it should be accepted until better grounds are found.[47a]

In the discussion of optics and magnetism there is the usual tendency to disregard the ancient and medieval contributions. The uninformed reader desirous of learning something of the subject could glean no more than the names, Euclid, Ptolemy and Alhazen.[48] As to the actual achievements of these men little is said. Thus all one could learn about Euclid is that he wrote the elements of optics, and of Ptolemy that he amplified Euclid's work. It would have been more valuable to point out briefly that until the time of Ptolemy the Greeks had been familiar with the phenomenon of reflection only. Ptolemy carried the science further in that he measured the phenomena of refraction, and tabulated the angles of incidence and those of refraction for water and glass.[49] But he was not able to formulate the law of refraction. Alhazen, again, improved on Ptolemy's apparatus for measuring the angles of refraction in different media and did very valuable work in discovering the structure of the eye. Indeed his methods strongly resemble the methods of today.[50]

[47a] See article, "*Licht*", vol. xvii.

[48] See article, "*Catoptrica*" (vol. vi) and "*Dioptrica*" (vol. vii). In connection with the latter subject the encyclopedist fails to indicate Ptolemy's and Alhazen's contributions to this branch of physics. According to him Kepler was the first to make progress in the field, in that he demonstrated correctly the properties of polished lenses (geschliffenen Gläzer). See also article, "*Sehekunst*", vol. xxxvi, pp. 1310-16.

[49] Ptolemy observed that the irregularities in stellar configurations were due to atmospheric refraction. He therefore wished to know the law of refraction, that is the relation between the angle of incidence and the angle of refraction; but he was unable to find any constancy in the relationship of these two angles.

[50] Hart, I. B., *The Great Physicists* (London, 1921), p. 20.

And yet all we get in the *Lexicon* is that Alhazen and Witelo wrote books on optics that are too difficult and prolix for beginners. It would seem as if none of these nine specialists was well enough versed in physics or any of its branches to have a firsthand acquaintance with the works in that field.

As for magnetism [51] it is asserted that the scholastics did not go any further than asserting that the magnet's attraction was due to a natural sympathy of the metal for the magnet. But an acquaintance with the work of Petrus Peregrinus (1269) would have convinced the lexicographer that somewhat more was known. We may say that in his *Epistola* he summed up all the knowledge of magnetism available at that time (1269). He made an experimental study of the magnet. He showed how to locate two points at which magnetic meridians of the sphere meet, and proposed the names north and south poles. He was aware of the fact that magnetization could be destroyed, and knew of the process of magnetization by induction.[52] Here firsthand knowledge would have been helpful to the encyclopedist.

The empirical tendency we noted in the physics of the eighteenth century is, not at all to our surprise, still to be found in chemistry too, for as a pure science it was in its development far behind physics, despite the work of Boyle who endeavored to separate it from medicine and alchemy. Chemistry in the *Lexicon* is still defined as " an art which is concerned with the breaking down of natural substances, animal, vegetable or mineral, by means of fire and other manipulations; with their recombination into whole substances; and with the preparation of medicine." [53] The very terms used in describing the methods of procedure are still

---

[51] See article, "*Magnet*" and "*Magnetismus*", vol. xix.

[52] Peregrinus, Petrus, *Epistle to Sygerus of Fonca Court* (trans. into English by Sylvannus Thompson, 1902, 30 pp.).

[53] "*Scheide Kunst*", vol. xxxiv, pp. 1109-1110.

those of Geber and the Alchemists. Thus "the whole science consists of solution (*solutio*) and crystallization (*coagulatio*); the former takes place through calcination and extraction; calcination through corrosion and ignition; these processes in turn take place through amalgamation, precipitation, stratification, fumigation, reverberation and dessication. The latter (i. e. *coagulatio*) occurs after exhalation, coction, congelation and fixation."[54] The only difference between chemistry (*Scheidekunst*) and alchemy is, as explained in the *Lexicon*, that chemistry is related to medicine and alchemy is concerned with turning baser metals into gold.

Turning to origins again, the lexicographer feels that the first chemist was Tubal Cain and not Hermes Trismegistos, not because he considers Hermes an Egyptian God, but because it was more likely that chemistry was brought to Europe by the Arabs and Greeks than by the Egyptians. But nothing is said about the Arabic contributions except that Avicenna and Averroes were opponents of alchemy, although Meyer states that both of these men followed in the footsteps of the alchemist Geber[55] who is omitted in this account. And Geber, by the way, or pseudo-Geber (thirteenth century, according to Berthelot)[56] contributed a considerable amount of chemical information such as recipes for certain preparations and the use of certain apparatus such as the water bath, ash bath and furnace. Paracelsus (1493-1541), again, is dismissed by the phrase that he obscured the art more than anything else, and therefore the *Lexicon* informs us many suspected him of having had a compact with the devil. And yet Paracelsus advocated "experimentation controlled by

[54] *Ibid.*, vol. xxxiv, p. 1110.
[55] Meyer, Ernst, *Geschichte der Chemie*, p. 29.
[56] *Ibid.*, p. 28.

authoritative literature ".[57] He rejected Aristotle, Galen, the four humors and advocated chemical therapy, which alone would mark him as a man of distinction. But tradition has viewed him as a blackguard and a quack, and so does the *Lexicon*. As for Boyle, his *Sceptical Chymist* is cited, in which he refutes both Aristotle's four humors and Paracelsus's three elements; and in another connection Boyle [58] is credited with having done much for experimental philosophy and chemistry. But a firsthand acquaintance with Boyle's works would have taught the lexicographer what Boyle wished to do for chemistry. He would have learned that Boyle's chief concern was to separate chemistry from medicine and alchemy. Although the lexicographer does separate chemistry from alchemy [59] he fails to see that chemistry as a science must divorce itself from all other empirical tendencies. This important part of Boyle's contribution seems to have escaped the eighteenth century encyclopedists.

To contrast with this thousand word article on chemistry there is one of ten thousand words on the philosopher's stone [60] and another of fifteen hundred words on alchemy.[61] In the opinion of the lexicographer there is much evidence for the belief in the transmutation of metals. And why should it not be possible, he asks, to multiply the seed of the metal and mineral kingdom, much like the seed of plants and animals? (There seems to be no departure from the alchemists' notion that metals grow like plants and animals). But it would be a mistake to suppose that metals could be per-

[57] Garrison, F. H., *Introduction to History of Medicine*, p. 206.

[58] See article, " Boyle ", vol. iv.

[59] In his definition of *Scheide Kunst* (vol. xxxiv, p. 1109) he says: " Sie wird Alchymia oder alchymistische Kunst Genannt."

[60] See article, " *Stein der Weisen* ", vol. xxxix. In this long article the belief is expressed that although no one has discovered it, the philosopher's stone does exist. (Vol. xxxix, pp. 1566-1567.)

[61] See article, " *Alchymie* ", vol. i, pp. 1065-1068.

fected outside their own kingdom. Nature teaches us that everything multiplies through its own kind, and the philosophers show that metals are improved only through metals. And just as there are two perfect metals, gold and silver, so there are two perfect tinctures, one white, the other red. In gold there is a golden seed from which the gold develops, in silver the silver seed. And if one could find the "philosophical mercury", then one might use it to grow and multiply.[62] Although it is admitted that there are many opponents of alchemy, there have been many who have held it in great esteem.[63] After citing the arguments for both sides, the editor concludes that although there are many bunglers in the profession, the whole art must not be condemned. Not only does he admit the possibilities of the art through the ordinary means of experimentation but he even succumbs, as a result of his religious bias, to the mystical side of alchemy. He does not seem to question the assertion that no one can be an adept except by careful guidance and instruction by word of mouth or through immediate enlightenment by God, that is by revelation. It is well to note the general tendency in the *Lexicon* to accept the method of science and at the same time to retain the method of mysticism in arriving at knowledge.

The relation of chemistry to medicine is well brought out in such articles as "*Secretum Secretorum*" of Paracelsus [64]

---

[62] See article, "*Stein der Weisen*", vol. xxxix, p. 1568.

[63] See article, "*Alchymie*", vol. i, p. 1067. The supporters, he says in one place, argue from experience, history and circumstance; thus in the year 1693 a goldsmith by the name of Gustenhoffer of Strassburg transmuted lead to gold before a varied audience. "Although he freely admitted", adds the encyclopedist, "that he did not know how to perfect the philosopher's stone, but that he had received it as a gift from a little man (männgen) whom he had given shelter during uncertain weather, the existence and possibility of this art is sufficiently proven."

[64] See article, "*Secretum Secretorum*", vol. xxxvi, pp. 931-934.

and in the many lengthy articles on sympathetic salts and salves.[65] What is the secret of secrets? It is a combination of certain chemicals ground in a mortar and then set in a sealed glass jar to warm in the sun or on the stove from twelve to fourteen days. Then it is a good cure for many diseases. If for example, it is used every morning as a salve, for that day one is insured against all contagious diseases, for it allows nothing to penetrate the skin, whether internally or externally. It also dispels melancholy, fantasy, loss of courage, heavy spirits; it is a cure for epilepsy, paralysis, hemorrhages in the head and so on ad infinitum. In another passage it is explained that the *Secretum Secretorum* may be prepared from a man's urine. If one administers this salt to a dog, then the disease of the person whose urine had been used is transmitted to the dog, and the patient is cured. In such fashion one can be freed without much trouble of the most desperate diseases. Then follow about fifty pages of description of sympathetic salts and salves that have effected cures, and for which the lexicographer attempts to give a naturalistic explanation. The chief peculiarity of these sympathetic salves is that they need not be applied to the patient's wound, but to the weapon with which the wound was inflicted. Also there are special instructions for the preparation of the salve; the ingredients must be gathered while the moon is waning and when it is in the sign of Venus, and the salve must be prepared only when the sun is in Libra. Then the salve is to be applied to the weapon once a day, when the wound is large, otherwise every second or third day will suffice. Another precaution is that the weapon must be wrapped in clean linen (a napkin once used by a woman during menstruation, no matter how clean, will not do) and put away in a clean place which is not too warm. If dust or wind blows on it, the patient will feel the pain.

---

[65] See article, "*Sympathetische Salbe*", vol. xli, pp. 732-734.

Such and more nonsense continues for pages. And then the explanation. The effect of the salve is due to the identity of the animal spirits in the man with those in the blood remaining on the weapon. The salve is called sympathetic because it has a sympathy for the wound, or a feeling for it, so that when it is applied to the blood on the weapon, its healing power is also extended to the wound with the aid of the pure spirit. Not until the last of the fifty pages devoted to these remarkable cures does the lexicographer admit that some authorities consider these remedies a pure deception and wholly unnatural. But between those who deny and those who affirm their efficacy, he concludes: " we wish to take the middle course, neither to repudiate the cure nor to credit it with too much power. However when we are asked why and in what manner its effect takes place, we are compelled to confess that we are unable to comprehend it." [66] We would like to believe that the hacks who concocted this drivel had their tongues in their cheeks; yet it is difficult to to believe that they would go to the length of fifty pages in doing so, especially after they admit some of the supposed curative powers.

The foregoing account of chemical therapy may serve as an introduction to the treatment of medicine. But before proceeding directly with that subject, we might raise another question. To what extent does the *Lexicon* recognize the advance from Galen that had been made with the publication of Vesalius's *Humani Corporis Fabrica* in 1543? All the evidence seems to point to the conclusion that neither had Galen [66a] gone out of fashion in 1750, nor was there much

---

[66] *Ibid.*, p. 734, "Wir wollen die Mittelstrasse gehen u. diese Krafft nicht ganz u. gar zu verwerfen ihr aber nicht zu viel zuschreiben. Wenn wir aber gefregt werden: wie und auf was Art diese Würkung geschehet, so werden wir gezwungen uns damit zu entschuldigen dasz wir es nicht begreifen können."

[66a] See article, "*Galenus*", (*Claudius*), vol. x, pp. 106-109.

of an appreciation of the scientific work of Vesalius.[66b] There is nothing in the *Lexicon* to indicate that Vesalius's work was epoch-making; on the contrary the lexicographer feels that he went too far in his criticism of Galen. As for Galen there is evidence in the *Lexicon* to show that Galenic views were still strongly entrenched. Although it is stated in one place that Harvey's discovery of the circulation of the blood [67] was a blow to Galen's physiological theory of the four humors, it is asserted in another article [68] that the brain contains animal spirits, and that health depends on the harmony of the humors.[69] The Galenic pulse doctrine is still retained, despite the work of Sir John Floyer (1669). In short, despite the fact that Harvey's discovery led to the growth of a iatro-physical school in medicine, the great majority of physicians still remained loyal followers of Galen.

Turning now to the question of cause and cure of disease, we may repeat the view of the *Lexicon* that although belief in stellar influence on human beings must be repudiated as a superstition, the sun and moon do influence disease, particularly diseases of the head. Apoplexy,[71] for example, is

[66b] See article, "*Vesalius*", vol. xlviii. Garrison, *op. cit.*, p. 219, notes that Vesalius' *Fabrica* "completely disposed of Galen's osteology and muscular anatomy for all time and recreated the whole gross anatomy of the human body."

[67] See article, "*Humores*", vol. xiii, p. 1172.

[68] See article, "*Gehirn*", vol. x, p. 615, "Das Hirn ist die Werckstatt derer Lebensgeister durch welche die Bewegung des Gantzen Cörpers geschiehet."

[69] See article, "*Pulse*", vol. xxix, p. 1234.

[70] See article, "*Gesundheit*", vol. x, p. 1335.

[71] See article, "*Gehirn*", vol. x, p. 615. "Unter denen Planeten wird dem Hirne der Mond und unter denen Metallen das Silber zugeeignet, daher die Haupt-Kranckheiten sich gerne nach des Mondes-Lauff richten und die aus Silber bereiteten Artzneyen vornemlich dem Haupte und Hirne dienen.

subject especially to lunar influence; consequently remedies containing silver are preferred. Then again, although amulets [72] are denounced as a superstition in one place, in another place they are admitted as cures. Nightmare [73] which is caused by a disturbance of the animal spirits may be cured by an amulet consisting of the root of peonies, of a wolf's tooth and of certain precious stones (chrysolyth, agate and emerald). An alternative cure is covering oneself with the skin of an ass. Epilepsy [74] which is also caused by a disturbance of the animal spirits, may also be cured by an amulet of the same ingredients laid on the spine. Although the absurd cures for hydrophobia, cited in Chambers, are omitted here, the absurd symptoms are still retained.[75] Those afflicted, it is claimed, bark and bite like dogs and even walk on all fours, not to mention the fact that their olfactory senses become as keenly developed as in the best of hunting dogs. What is more, those stricken with the disease actually believe themselves transformed into the canine beast. And when the disease results from the bite of a mad dog, they are thrown into a violent rage and fever, as a result of which they usually succumb. In answer to the question as to the cause of all this disturbance, the Galenic lexicographer replies that it is due to a strange movement of the animal spirits that gets beyond the control of the soul. As for treatment, Lémery comes to our aid with the suggestion that if the patient is securely bound and thrown into a cold bath, it might help; but then there is the danger that the person attending the patient might be bitten. Phlebotomy is another suggested cure.

[72] See article, "*Amuletum*", vol. i, pp. 1818-19. A spirited repudiation of its supposed efficacy.

[73] See article, "*Alp*" (nightmare), vol. i, pp. 1327, 1328.

[74] See article, "*Epilepsia*" (*morbus caducus*), vol. viii, p. 1342.

[75] See article, "*Hydrophobia*" (*wasserscheu*), vol. xiii, pp. 1377-1381.

This leads us to turn to the article on phlebotomy.[76] It does not seem to have lost any of its popularity. An interesting bit of "information" which might not be found elsewhere is the story of how man discovered this cure. The fact is that man has learned it from the dumb beast. The Hungarian horse and an American animal called Dante (*mirabile dictu*) were accustomed to cut their veins with their teeth to get rid of surplus blood. Although, the *Lexicon* admits, not all physicians are agreed as to the therapeutic value of phlebotomy, daily experience proves that it must not be put aside. In certain cases it should be used since it does effect a cure. An interesting precaution is added. Unless the need is immediate, it is best to perform the operation in the autumn or spring, and according to Stahl at the time of the equinox.

Another type of remedy favored throughout the middle ages is the use of parts of animals. These remedies still retain their old values in the *Lexicon*, quite undiminished. Thus the ape's[77] heart roasted, dried and pulverized strengthens the heart and dispels melancholy. If the heart of the ape is placed under a man's head while asleep, he will see strange and frightful dreams.[78] The ashes of the stork[79] distilled into an essence are used against poisons and the pest; while its excrement dissolved in water serves as a cure for epilepsy. The brain of a dog[80] (he must be of a solid color), when cooked and eaten, dispels delirium that was caused by the bite of a mad dog; and the gall of a black suckling dog ground into a powder or mixed with vinegar is a cure for epilepsy. Dog's teeth burnt to ashes, mixed with honey, and

---

[76] See article, "*Aderlass*", vol. i, pp. 493-94.
[77] See article, "*Affen*" (Apes), vol. i, p. 717.
[78] The authority cited for this strange effect is Rhazes.
[79] See article, "*Storch*", vol. xl, p. 427.
[80] See article, "*Hund*", vol. xiii, pp. 1188-89.

smeared over children's jaws will facilitate teething. In diseases of the brain, furthermore, a young dog when opened and laid still warm on the head, will effect a cure.[81] Similarly the different parts of the wolf have their therapeutic values.[82] Needless to say there is no abatement of theriacs and electuaries;[83] there are pages upon pages full of recipes.

The *Lexicon* takes issue with Descartes on the subjects of human physiology and animal psychology. In the first place the pineal gland is not the seat of the soul, as Descartes would have it.[84] Then as regards animals, it again contradicts Descartes, and credits them with immortal souls, reason and speech.[85] The argument is somewhat along the following lines. The mistake Descartes made was to identify the essence of the soul with thought, and thus to reject the sensitive soul. Hence he considered animals machines. But if the existence of a sensitive soul is admitted, it cannot be

[81] In the article on the cat (vol. xv) we find similar credulities. The ear of the living cat expels the worm from the finger and prevents its recurrence, if the patient will insert his finger in the cat's ear a few times a day and keep it there for about fifteen minutes. To cure a pain in the side, open up a living cat and apply it to the affected part. If one drinks three drops of blood from the vein of a cat's tail, mixed with water, one is immediately cured of the *Bösewesen* (evil spirit).

If you burn the head of a black cat to ashes and then blow some of it three times a day into the eyes, it is effective enough to expel a cataract. Some people always carry with them the placenta of a black cat as a remedy against any kind of eye trouble. And wanton maidens sometimes make a love potion from the cat's brain.

[82] See article, "*Wolf*", vol. 58. As for the lion (vol. xviii) if one anoints oneself with the fat from the lion's kidneys, wolves will be frightened away; and toothaches can be easily prevented by merely hanging a lion's tooth about the neck.

[83] See article, "*Electuarium*" *advisum, Electuarium vital, etc.*, vol. viii.

[84] See article, "*Gehirn*", vol. x, p. 615, "glandula pinealis...ist einmahls wie wohl vergeblich, von dem Cartesis vor dem Sitz der Seele gehalten worden."

[85] See article, "*Thier*", vol. xliii, pp. 1333-1382 about 25000 words.

denied that the animal's brains and sense organs would be affected by external stimuli as are those of human beings. The lexicographer, however, is not satisfied with proving that the animal feels. He insists that the animal reasons too. The argument here is not so convincing, although it seems simple and almost self-evident to him. "Who can doubt", he asks, " that animals reason when they deceive man himself?"

Arguments of a similar nature are then adduced to prove the animal's soul immortal. The fact is, it is asserted, that

not only individuals but whole nations such as the Egyptians and the Persians believed in the immortality of the animal's soul. Surely they must have found something in the soul of the animal which seemed to them so noble that they could not imagine it to be destined to decay and disappearance. This has remained the opinion to this day.

But the strongest argument for the immortality of the animal's soul is this: That which has been brought forth by God is part of God's purpose and therefore immortal. That animals can talk is proved in the *Acta Eruditorum* of 1736 (p. 287) where Leibnitz is cited as having known a dog that was able to pronounce more than thirty words, and to answer a number of its master's questions. Reliance on authority, therefore, and the belief in creation as a revelation of divine omniscience are still much in the way of a scientific point of view.

Animals are not only endowed with speech and reason, but with an uncanny ability for weather forecasting. The farmer, for instance, can be certain of rain when the cow pants for breath toward noon; (2) when the pigs pull the straw hither and thither, throwing it about as if they were mad; (3) when dogs eat grass and then throw it up; (4) when they roll on the ground and don't eat at all, but scratch the earth and bark in the morning; (5) when the cats lick their bodies and stroke their ears with their paws; (6) when oxen lick their feet and run bellowing to the stable; (7) when the calves run hither and thither under the horses or other large animals, or run about as if they were mad; (8) when the cows dig into the ground with their feet or horns; (9) when the asses or mules shake their ears in an unusual way; (10) when the goats hasten with immoderate speed to their fodder and will not be driven from it even when beaten; (11) when the sheep on their way home eat the grass along the road, and will not be driven from it; (12) when the deer buck and fight with one another; (13) when wolves and foxes howl and come near the villages; (14) when the eagle stops up the holes in its nest.

Although the progress of botany [86] in the seventeenth century is summarized in the *Lexicon,* its medieval lore has not been abandoned in the work. Thus it is defined as a science that is concerned not only with the classification of herbs, plants and flowers, but also with their powers and effects. And a botanist is defined as one who can distinguish and classify the nature and virtue of plants. Botanists, it is added, can be classified into two groups, those like Bauhin and Cesalpinus who were interested in classification, and those like Carrichter and Paracelsus who were more concerned with the therapeutic values of plants. Carrichter, it may be noted in passing, still classified plants according to the twelve signs of the zodiac.[87]

From the above definition it becomes evident that the eighteenth century had not yet been able to separate botany from medicine. Nor for that matter had it as yet rid itself completely from the occult virtues of plants. Although the efficacy of amulets is repudiated in one article,[88] it is asserted in others that the mistletoe has the power of preventing

---

[86] See article, "*Botanica*", vol. iv.
[87] Meyer, *Geschichte der Botanik,* p. 433.
[88] See article, "*Amuletum*", vol. i, p. 1819.

epilepsy, when it is hung around the neck, and nosebleed when only held in the hand.[89] The acanthus root again is an effective cure for the pestilence when worn as an amulet on the finger.[90] With regard to the meadow-saffron [91] some medical authorities deny and others affirm its occult powers to resist pestilence and other poisonous diseases. Occult powers are repudiated in an article on the subject;[92] yet the lime tree [93] seems to have strange curative powers. An interesting case is that of an insane young man in a rage. The strongest men were unable to check him. But when his hands and feet were tied with lime bast, he was immediately put to rest. Nor are the virtues of *agnus castus* repudiated.[93a]

The treatment of precious stones may be characterized as full of contradictions. In the short article on the subject [94] their use in medicine is definitely repudiated as a superstition; yet the medicinal virtue of the agate [95] and the amethyst [96] are not denied, nor are the occult qualities of

[89] See article, "*Eichen—Mistel*", vol. viii, p. 460; also article, "*Mistel*", vol. xxi, p. 516. Albertus Magnus is still cited as an authority for the notion that the mistletoe is a preventive medicine against the pestilence, if taken internally morning and evening.

[90] See article, "*Acanthus*", vol. i, pp. 252, "Einige bereiten aus dieser Wurtzel, in gewissen Zeichen gegraben, einen Ring u. tragen denselben an den Gold Finger als ein Amuletum zur zeit der Pest." There is no effort here to repudiate its efficacy.

[91] See article, "*Bulbus pratensis*", vol. iv, p. 1908.

[92] See article, "*Verborgene Eigenschafft*", vol. xlvii, pp. 186 *et seq.*

[93] See article, "*Lindenbaum*", vol. xvii.

[93a] See article, "*Agnus Castus*", vol. i.

[94] See article, "*Edelgesteine*", vol. viii.

[95] See article, "*Achates*", vol. i. Its virtue as an antidote is denied but its power to "still the blood" is not denied (p. 316).

[96] See article, "*Amethystus*", vol. i, p. 1728. Some of the reputed virtues are repudiated as imaginary, e. g., as a cure for drunkenness when placed on the navel, worn on a finger or taken internally. But it is not denied that it serves to check dysentery, and to cure acidity in the stomach since it is an alkali.

others. Thus if the alectory [97] is held in the mouth, it has the solar power of making one valiant and victorious. The blood stone [98] (*lapis haematites*) if held in the hand stops excessive nose bleeding, and the *lapis lazuli* [99] when worn on the hand purifies the blood and dispels melancholy and fantasy. Since the *lapis variolae* [100] is effective in drawing out poison, it is a cure for measles when hung about the necks of children, so that it covers the heart. Then there is the *lapis hystericus* [101] that seems to have a strange power. It is rather a peculiar stone. On one side it is curved. On its flat surface can be discerned both the male and female genital organs. One writer therefore concludes that it is an effective cure for hysterics (*mutterschwachheit*), and against asphyxiation. It might also be helpful if either the male's power or the female's fruitfulness had been bewitched. At the same time the virtues of other stones are condemned as purely imaginary and the product of superstition. The belief, for instance, that the *lapis stellaris* [102] can resist the pest and witchcraft is denounced as superstition. Similarly the power of the agate to withstand the poison of the scorpion is also denied; and certain powers of the amethyst are frowned upon, such as its effectiveness as a cure for drunkenness, when placed on the naval or worn on the finger.

In the treatment of the occult sciences, the *Lexicon* can hardly maintain a sceptical attitude when it admits belief in a devil, witchcraft, magic and the like. The devil is described as a "spiritual substance endowed with will,

[97] See article, "*Lapis Alectorius*", vol. xvi, "Diesen Stein im Munde gehalten soll wegen seiner solarischen Krafft streitbar u. siegbar machen."

[98] See article, "*Lapis Haematites*", vol. xvi.

[99] See article, "*Lapis Lazuli*", vol. xvi.

[100] See article, "*Lapis Variolae*", vol. xvi.

[101] See article, "*Lapis Hystericus*", vol. xvi.

[102] See article, "*Lapis Stellaris*", vol. xvi.

reason and limited power to produce effects." [103] He plays, for instance, a very important part in the practice of magical ligatures. Although this practice, used to deprive males of their potency, is denounced as a superstition, its effects are admitted, and its cause attributed to the devil. Says the lexicographer: "Every one must admit that the devil in such cases does his best. For otherwise it is impossible to fathom what ligatures and other ceremonies are supposed to effect." [104] But besides the devil, witchcraft may also produce male impotence.[105] Such a case is very difficult to cure. Yet Van Helmont had a suggestion which the lexicographer feels would be very helpful to the reader. The afflicted one may be cured if he steps over a broom before he urinates. Another remedy is to boil the patient's urine in a tightly closed pot. The witch that cast the spell will under such circumstances become alarmed, will request to have the pot removed and will promise to lift the spell. Still another suggestion is for the man to pass his urine through his wife's wedding ring.

Although superstition [106] is identified with irrational attitudes, with the tendency to attribute supernatural powers to stones, plants and herbs, and with the belief of the ignorant in astrology, we have seen that this correct interpretation of the subject did not prevent the editor from accepting a good many notions that can only be characterized as superstitions. And in other places we find other evidence of the retention of specific superstitions despite the repudiation of superstition in general. Thus the devil's existence is admitted. As for witchcraft, the editor here assumes the rôle of a neutral.[107]

[103] See article, "*Teufel*", vol. lxii.
[104] See article, "*Nestelknüpfen*", vol. xxiii, p. 1959.
[105] See article, "*Penis erectio laesa*", vol. xxvii.
[106] See article, "*Aberglaube*", vol. i, pp. 107-111.
[107] See article, "*Hexerey*", vol. xii, p. 1979.

Inasmuch as the learned are not in agreement on the question, he refuses to take sides between those who cite scripture and the testimony of witches in support of the art, and those who deny both the witches' compact with the devil and the possibility of a bodily union with him. And yet in the discussion of black magic [107a] he asserts that there is nothing more foolish than to attempt to produce an unnatural effect with the aid of the devil. In another connection, we learn that witches can do bodily harm to human beings.[107b] By means of a ligature they can deprive the male of his potency, and not entirely without the aid of the devil. Nevertheless the power of the devil is limited. He cannot transform human beings into beasts. It is also conceded that there is not enough evidence to establish the witches' sabbath as a fact.[107c]

Necromancy, the magical art of calling up the dead through the intervention of the devil, and exacting secrets, is not repudiated. But the devil does not really call up the dead, only the image of the dead, as in the case of Saul and the witch of Endor. Saul saw the image of Samuel produced by the devil. Although the witch's compact with the devil is denied, in this biblical case one might suspect that there was some collusion between the witch of Endor and the devil.[108]

If black magic, as we have seen above, is pronounced foolish,[109] natural magic [110] is approved as the means of producing strange and singular effects through natural but still secret powers. Some of these strange effects are the pro-

---

[107a] See article, "*Magie*", vol. xix, p. 297.
[107b] See article, "*Nestelknüpfen*", vol. xxiii.
[107c] See article, "*Hexerey*", vol. xii.
[108] See article, "*Necromantie*", vol. xxiii, p. 1538.
[109] See article, "*Magie*" (die Schwarze) Magia Diabolica, vol. xix.
[110] See article, "*Magie*", vol. xix.

duction of lice and mice and other little animals from putrid matter, the transplantation of disease, preparation of love potions, and even the production of impotence.[110a] By the same art certain marvelous transformations can be brought about. For example an ant placed between two saucers, will, on the third day after being buried in earth, be transformed into a snake. Or if a bullock's mouth and nostrils are stuffed up and it is beaten to death with a stick, a swarm of bees will emerge from the carcase. If the marrow of a man's bone is burned in rubbish together with the hair of a female that menstruated, snakes will arise from the union. These and other almost miraculous effects are produced because of the secret powers of nature and also because of man's peculiar dexterity.[111]

The corpuscular theory, according to one authority cited in the *Lexicon*, would seem to throw some light on how strange effects are produced by natural magic. The following seven propositions seem to the author to clarify the matter somewhat: (1) Everything is full of light. (2) Reflected light carries the atoms back with it. (3) The air as it circulates in the earth is full of dust particles. (4) Natural bodies can be divided indefinitely. (5) The atoms, contrary to the common laws of motion, are attracted to one another. (6) Like dust particles attract one another. (7) The little bodies are composed of spirit and matter, (*Geist und Materia*). From this last proposition is to be derived the whole basis of magic. Whenever the " spirits " which cause everywhere very swift motion are directed artificially to a

---

[110a] See article, "*Zauberey*", vol. 61, p. 77: "denn man kan oftmals durch die natürliche Magie gantz seltsame u. erstaunende Dinge ausrichten, und bezeugen die jenigen welche davon geschrieben, wie es gantz natürlich zugehe dasz man Läuse, Mäuse u. andere dergleichen kleine Thierchen herfürbringen, die Mannheit binden, die krankheit auf eine seltseme weise Fortpflantzen, Liebes Trancke verfertigen könne."

[111] See article, "*Magie*" (Die Curiose), vol. xix, pp. 298-299.

particular goal, they can then exercise their power and produce astounding effects. This "explanation", it is acknowledged, does not explain many effects produced by natural magic which can only be ascribed to the strange and secret powers of natural bodies.[112]

If we pass on from natural magic to magic words,[113] we find that they too have not entirely lost their effectiveness. The lexicographer admits it is difficult to determine whether written or spoken words have the power of curing patients in a natural way, yet there are many instances where such cures have been produced. There is the case of the patient whose head was filled with worms. When all other remedies failed, he was cured only after he had spoken certain words as he broke off certain leaves. Another case is that of a man who suffered from an incurable hemorrhage. He was given an unknown word inserted in a raisin. When he swallowed it he was cured. Still another case is that of a young man suffering from consumption who was finally cured by words.

In the article on the divining rod [114] we find the statement that the mountain folk used it to discover copper mines, because the wood of which it is made has an affinity for the metal. The superstitious ceremonies, adds the lexicographer, can be dispensed with.

The art of divining the disposition and fortune of a person from the lines of his hands, that is, chiromancy,[115] still finds support in this *Lexicon,* more, indeed, than it found in Chambers. To be sure a moderate position is taken. Although the extreme position that the whole art is infallible is untenable, it is well to recognize the purpose of the lines and to

---

[112] See article, "*Magie Naturalis*", vol. xix, pp. 300-301.
[113] See article, "*Magische Worte*", vol. xix, pp. 316-317.
[114] See article, "*Wünschelrute*", vol. 59.
[115] See article, "*Chiromantie*", vol. v.

explain through them bodily dispositions of man as well as his fortunes or misfortunes. Thus if a line is present under the small finger, it is certain that the person will be married at least once. Apparently the editors of the *Lexicon* accepted the notion prevailing at the time that there is a close relationship between the lines of the hand and the internal parts such as the heart and the liver upon which depend the passions and inclinations.

Fascination,[116] on the other hand, is discredited on the ground that it is based on a false theory of vision, namely that rays are emitted from the eye to the object. Since that is false, it is also wrong to imagine that one person can bewitch another by directing the rays of his eye upon him, since such rays are only a figment of the imagination.

However, physiognomy [117] is discredited on somewhat less reasonable grounds. Since, argues the lexicographer, two people with more or less identical dispositions will not necessarily have the same facial features, it is difficult to see how there can be any relationship between character or disposition and facial features.

[116] See article, "*Fascination*", vol. ix.
[117] See article, "*Physiognomia*", vol. xxvii.

CHAPTER IV

Conclusion

It is evident that we cannot measure with scientific precision the extent to which medieval science has survived in the two encyclopedias under consideration. But if the foregoing analysis is correct, we cannot avoid the conclusion that the scientific revolution of the seventeenth century did not effect a complete overthrow of the science and superstition of earlier centuries. Indeed the two encyclopedias still bear a close resemblance to their earlier medieval prototypes. The very classification of knowledge in Chambers' work, and incidentally this classification is retained by the far more sceptical *Encyclopédie*,[117a] retains its medieval character, with its division of metaphysics into ontology and pneumatology; chemistry into alchemy and natural magic; with its optics, dioptrics and catoptrics. The deluge theory of fossils is not at all questioned, despite the scientific beginnings in the geological field made by Steno in 1669, when he published his *De Solido intra Solidum* (in which he explained that the earth's crust consisted of parallel layers and that fossils were remnants of organic matter), and the work of Leibnitz (*Protogea*, 1693) who explained the organic origin of fossils and asserted that such an explanation was not in opposition to the Bible.[118] Also the conception of the four elements and the phlogiston theory of fire are still in good repute.[119]

[117a] Thorndike, L., "L'Encyclopédie and the history of science," pp. 385-86.

[118] Ornstein, M., *The Rôle of Scientific Societies in the Seventeenth Century* (New York, 1913), p. 25.

[119] See article, "Fire" in Chambers.

CONCLUSION 75

It is true that in one or two minor details, the *Lexicon* evinces a greater scepticism than the *Cyclopedia*. We noted, for example, how the art of fascination and the pseudo-science of physiognomy are accepted and justified by Chambers, and repudiated in the German work. Also, in the discussion of the origin of fossils, the *Lexicon* goes so far as to explain the organic theory side by side with the deluge theory. Although the editor scouts the possibility of such a development, he is nevertheless constrained to cite one authority who contends that the flood theory alone is an inadequate explanation, since one finds petrified bodies in different layers. Hence, if fossils found not very deep in the ground are the result of one deluge, then fossils found deep down in the earth must be accounted for by other earlier floods.[120] On the whole, however, the *Lexicon* is no more scientific or sceptical than the *Cyclopedia*. On the contrary, in some respects it is less so, and exhibits even more of a medieval character. Despite its sincere efforts to appreciate scientific progress the *Lexicon* cannot wholly depart from the medieval notion that all the branches of knowledge are really the handmaidens of theology. This is suggested by the editor's enumeration of the different branches of knowledge,[121] where theology heads the list, followed in turn by jurisprudence, medicine and finally mathematics. In another connection this attitude is more definitely expressed. The editor there is concerned with a defense of the Pietists, who were attacked as enemies of science since they emphasized the superiority of revealed over empirical knowledge. The answer as given in the *Lexicon* is that Spener was not opposed to science at all. On the contrary he favored the study of all branches of knowledge, for all could perform a valiant service for theology. It is this theological bias

[120] See article, "*Sündfluth*", *Lexicon*, vol. lxi, p. 117.
[121] See *Vorrede in Lexicon*, vol. i, p. 6.

which weakens somewhat the treatment of science in the *Lexicon*.[122] It may be worth while recalling here how this attitude is expressed in the discussion of the purpose of the study of science, which is nothing more nor less than to combat atheism, and to prove the divine nature of things. And yet the editors of the *Lexicon* complained that Aristotle's science was based on metaphysical speculation and not on real principles. Obviously they did not realize that the speculative hypothesis that all things have a divine nature cannot lend itself to experimental verification nor can it yield any exact knowledge of the physical nature of things. Chambers, we must note, was free from such assumptions. Chambers was a free thinker. Hence we find also that he has very little to say about the devil, except to quote someone's *bon mot* that the Ethiopians paint the devil white to be even with the Europeans who paint him black.[123] The devil in the *Lexicon*[124] on the other hand, is quite real, although shorn of some of his powers; and the editors devote as much space to a discussion of his nature and powers as they do to the subject of comets (about ten thousand words).

Their orthodox religious position, again, prevents the lexicographers from evaluating fairly the scientific approach of such a man as Spinoza. Because he identified God with the universe and repudiated revelation, he is denounced as an atheist and as the patriarch of all atheists. An atheist and a Spinozist are synonymous terms.[125]

But even from another point of view, we can trace the umbilical cord of both works back to the middle ages. Although to-day we may see no resemblance between science and magic, " in their history ", says Thorndike, " science and

[122] See article, "*Wissenschaften*", vol. 57, p. 1462.
[123] See article, " Devil " in Chambers.
[124] See article, "*Teufel*" in *Lexicon*, vol. lxii.
[125] See article, "*Spinoza*", vol. lxii.

magic were not unassociated. Scientists might accept magical doctrines and magic might endeavor to classify its fancies and to account for them by natural causes." [126] And that is precisely what we find in both of these encyclopedias. Chambers' defense of witchcraft and physiognomy, and the *Lexicon's* explanation of the curious effects produced by natural magic are exactly efforts to give natural or scientific explanations to the most fanciful notions current in the eighteenth and earlier centuries. It is precisely, however, from this same point of view that we can see magic and science drifting further and further apart. This tendency too can be noted in both of these compilations. Just as some fancies are explained by natural causes, so others are discarded as chimerical and absurd. Hence we must conclude that the scientific tendency, although somewhat obscured by the persistence of occult science and superstition, is already there, gradually shedding its pseudoscientific skin, only not as rapidly as we might be led to believe from a casual examination of the results of scientific progress in the seventeenth century.

[126] Thorndike, L., *The Place of Magic in the Intellectual History of Europe* (New York, 1905), p. 35.

BIBLIOGRAPHY

Allbutt, T. C., *Greek Medicine* in Rome, London, 1921.
——, *Fitzpatrick Lectures on the History of Medicine* delivered in 1909-1910, and other essays, London, 1921.
——, *Historical Relations of Medicine and Surgery till the Sixteenth Century*, London, 1905.
D'Alembert, *Discours Préliminaire de L'Encyclopédie* with introduction and notes by L. Ducros, Paris, 1893.
Bancroft, W. D., *Methods of Research* (Rice Institute Pamphlet, vol. xv, Oct., 1928).
Barry, F., *The Scientific Habit of Thought*, New York, 1927.
Biedermann, K., *Deutschland im Achtzehnten Jahrhundert*, vol. ii, part 2 till 1740, Leipzig, 1880.
Boyle, Robert, *The Sceptical Chymist*, London, 1661, and vol. i, pp. 300-370, of *Works*, London, 1744.
——, *Specific Medicines*, London, 1685.
Buckley, A. B., *A Short History of Natural Science* (elementary), New York, 1888.
Buckley, H., *A Short History of Physics*, London, 1927.
Burtt, E. A., *The Metaphysical Foundations of Modern Physical Science*, New York, 1925.
Bury, J. B., *The Idea of Progress*, London, 1920.
——, *A History of the Freedom of Thought*, London, 1913.
Chambers, E., *Cyclopedia; or An Universal Dictionary of Arts and Sciences*, London, 1728, 2 volumes.
Crew, H., *The Rise of Modern Physics*, Baltimore, 1928.
Cru, R. L., *Diderot a Disciple of English Thought*, New York, 1913.
Cumston, C. G., *An Introduction to the History of Medicine to the End of the 18th century*, London, 1926.
Duhem, P., *Etudes sur Léonard de Vinci*, 3 vols., Paris, 1906-1913.
Garrison, F. H., *Introduction to History of Medicine*, Philadelphia and London, 1929.
Gerland, E., *Geschichte der Physik*, Munich, 1913.
Guhrauer, G. E., *G. W. von Leibnitz, eine Biographie*, Breslau, 1842.
Gregory, J. C., *A Short History of Atomism*, London, 1931.
Hart, I. B., *The Great Physicists*, London, 1921.
Haskins, C. H., *Studies in the History of Medieval Science*, Cambridge, 1924.

BIBLIOGRAPHY

Kopp, H., *Beiträge zur Geschichte der Chemie*, 3 vols., Brunswick, 1869-1875.
Lasswitz, K., *Geschichte der Atomistik vom Mittelalter bis Newton*, 2 vols., Hamburg, 1890.
Libby, W., *The History of Medicine*, Boston, 1922.
———, *An Introduction to the History of Science*, Boston, 1917.
Meyer, E. von, *Geschichte der Chemie*, Leipzig, 1914.
Ornstein, M., *The Rôle of the Scientific Societies in the Seventeenth Century*, New York, 1913.
Peregrinus, Petrus, *Epistle to Sygerus de Foncacourt*, translated by S. Thompson, 1902, 30 pp.
Robinson, H., *Bayle the Sceptic*, Columbia University Press, 1931.
Rossiter, W., *Dictionary of Scientific Terms*, London, 1878.
Sarton, G., *Introduction to the History of Science*, vol. i, Washington, D. C., 1927.
Singer, C. (editor), *Essays on the History of Medicine*, Zurich, 1924.
Smith, P., *A History of Modern Culture*, New York, 1928.
Thorndike, L., *The Place of Magic in the Intellectual History of Europe*, New York, 1905.
———, *Science and Thought in the Fifteenth Century*, New York, 1929.
———, *History of Magic and Experimental Science*, 2 vols., New York, 1923.
———, " *L'Encyclopédie* and the History of Science" in *Isis*, VI (1924).
Whewell, W., *History of Inductive Sciences*, 3 vols., London, 1837.
Wolf, R., *Geschichte der Astronomie*, Munich, 1877.
Zedler, J., *Grosses Vollständiges Universal Lexicon aller Wissenschaften und Künste*, Halle and Leipzig, 1732-1750, 64 volumes.

INDEX

Adelard of Bath, 15, 50
Albertus Magnus, 47, 48, 49
Alchemy, 57, 58
Alembert, Jean D', 12, 13, 35
Alfonsine Tables, 40
Alhazen, 18, 19, 54, 55
Al-Khwarismi, 14
Almagest, 39
Amulets, efficacy of, 31, 66, 67
Arnald of Villanova, 26
Aristotle, 21, 45, 46, 51, 57
Astrology, treatment of, 16-17, 42-44
Astronomy, treatment of, 14-15, 39-42
Atomic Theory, treatment of, 49-51
Averroes, 56
Avicenna, 56

Bacon, Francis, 45, 53, 54
Bacon, Roger, 38, 45, 47, 48, 49, 51
Bartholomew of England, 48
Bauhin, 66
Bayle, Pierre, 8, 43
Boerhaave, 27
Botany, treatment of, 22-23, 66-67
Boyle, Robert, 16, 20, 22, 31, 50, 51, 52, 55, 57
Brain, treatment of, 25-26
Buridan, Jean, 49
Bury, John, 24, 47

Capella, Martianus, 40
Carrichter, 66
Cartesianism, 24
Cesalpinus, 66
Chemistry, treatment of, 19-21, 56, 57
Chiromancy, treatment of, 33, 72-73
Corpuscular Theory, 18, 53
Comets, influence of, 17-18, 41
Copernicus, 40, 41, 42, 43
Corneille, Thomas, Le Dictionnaire des Arts et des Sciences, 8

Dalton, John, 51

Descartes, René, 25, 64
Devil, treatment of, 68-69, 76
Diderot, Denis, 12, 13, 35, 37, 42
Disease, treatment of, 27-30
Divination, treatment of, 32-33, 72
Du Hamel, J. B., 49
Duhem, Pierre, 49

Epicurus, 49
Epilepsy, cure for, 62, 66-67
Euclid, 54
Experimental method, 19, 52-53

Fossils, deluge theory of, 74, 75
Fracastoro, Geronimo, 51
Furetière, dictionary of arts and science, 8

Galen, 27, 60, 61
Galileo, 43, 49, 52
Gassendi, 49
Geber, 56
Gems, therapeutic value of, 23, 24, 30, 67-78
Gerard of Cremona, 15
Grosseteste, Robert, 47

Harvey, William, 52, 61
Haskins, C. H., 15
History, definition and philosophy of, 13, 14, 37-38
Hydrophobia, cause and cure, 28-30, 62

Isadore of Seville, 16

Jablonski, J. T., *Allgemeines Lexicon der Wissenschaften u. Künste*, 8

Kepler, J., 41, 43, 52

Leibnitz, G. W., 7, 8, 19, 74
Ligatures, 69
Lémery, N., 62
Louis XIV, 7

81

INDEX

Lucretius, 49

Magic, treatment of, 30, 32, 70, 71, 72
Magnetism, treatment of, 54, 55
Medicine, history of, 26-27; treatment of, 25-30, 60-64
Messahalla, 15, 39
Moréri, *Grand Dictionnaire*, 8

Necromancy, treatment of, 33, 70
Newton, Isaac, 52, 53
Nicholas of Cusa, 40, 51

Occam, William of, 49
Occult Virtue, 66-67
Optics, treatment of, 54, 55
Oresme, Nicholas, 40

Paracelsus, 20, 21, 27, 43, 56, 66
Peckham, John, 48
Peregrinus, Petrus, 55
Peurbach, George, 39
Peter of Abano, 26
Philosopher's Stone, 57
Philosophy, natural, treatment of, 44-49
Philtres, efficacy of, 31-32
Phlebotomy, 28, 63
Phlogiston, 74
Physics, 18-19

Physiognomy, treatment of, 33-34, 73, 75, 77
Pico della Mirandola, 43
Psychology, animal, 64-66
Ptolemaic system, 14, 40
Ptolemy, 14, 42, 54
Pythagoras, 14, 39

Regiomontanus, 39

Sacrobosco, J., 14
Secretum Secretorum, 58-60
Spener, 75
Spinoza, 76
Steno, 74
Sydenham, 27

Thorndike, L., 48, 76-77
Tycho Brahe, 40, 41

Undulatory theory, 18, 53
Universal solvent, 22

Vesalius, 60, 61
Vinci, Leonardo da, 40

William of Conches, 50
Witchcraft, treatment of, 33, 69-70, 77
Witelo, 18, 19, 55

Zacuto, Abraham, 40

Bei Fragen zur Produktsicherheit wenden Sie sich bitte an:
If you have any questions regarding product safety,
please contact:

Walter de Gruyter GmbH
Genthiner Straße 13
10785 Berlin
productsafety@degruyterbrill.com